P. MICHAEL LEE

THE ARMCHAIR
AVIATOR

THE ARMCHAIR LIBRARY

THE ARMCHAIR QUARTERBACK
THE ARMCHAIR AVIATOR

THE ARMCHAIR LIBRARY

THE ARMCHAIR
AVIATOR

Edited by John Thorn
with David Reuther
Illustrations by Bob Carroll

CHARLES SCRIBNER'S SONS
New York

Permissions appear on page 308.

Library of Congress Cataloging in Publication Data

Main entry under title:

The armchair aviator.

1. Aeronautics—Addresses, essays, lectures.
I. Thorn, John, 1947–   . II. Reuther, David.
TL559.A75   1984   629.13   83-20051
ISBN 0-684-18014-6

1 3 5 7 9 11 13 15 17 19   F/C   20 18 16 14 12 10 8 6 4 2

Printed in the United States of America.

CONTENTS

Introduction

We have always known what we wanted. The dream and the drive were old when the story of Icarus was new: to scale the heavens, to attain the serenity of the birds and the wisdom of the angels, to glimpse eternity and not know death. To be as free as the wind. To fly.

Once realized, the dream of flight could have been expected to lose much of its hold, but it has not, for the dream was from the outset more spiritual than physical. The urge to fly is not merely a desire to flirt with death (although surely that element exists, as it does for such determined risk-takers as mountaineers, deep-sea divers, and race-car drivers); it is an urge to know, to feel, to be, in a way that is impossible with one's feet on the ground.

"The swallows fly over our head laughing at us," wrote a frustrated Otto Lilienthal in 1895, sensing how near man must be to mastering the firmament and yet how far. The progress in flying machines from Lilienthal's glider, in which he hurtled to his death in 1896, to the Wright Flyer of only seven years later was not a small evolutionary step but a giant leap, on the order of the rise from reptile to bird. Another giant leap, this one described by Neil Armstrong, was the moon landing of 1969. What will be the next great advance? Now that man has navigated air and space mechanically, are we, as Charles Lindbergh speculated in his final year, "approaching a stage in evolution when we can discover how to separate ourselves entirely from earthly life, to abandon our physical frameworks in order to extend both inwardly and outwardly through limitless dimensions of awareness? In future universal explorations, may we have no need for vehicles or matter?''

Perhaps—and if so, man would come full circle to his earliest yearning. Even for Lindbergh, a man of singular accomplishment, the primordial dream of spirit ascending fired his imagination to the end.

The saga of man in the air, from dream to reality, from aeronaut to astronaut, is what this book is about. The editor's aim has been to provide, through the best that has been written on the subject, a panoramic view of aviation's history, lore, technique, literature, and, above all, people.

You don't have to be a flier to enjoy such writers as James Michener, Roald Dahl, or Joseph Heller—nor even such ''aviation writers'' as Wolfgang Langewiesche or Harald Penrose. (Was Thoreau a ''nature writer''?) You don't have to be a flier to be moved by ''Blossoms in the Wind'' (the last letters home of four kamikaze pilots), to smile at James Thurber's ''The Greatest Man in the World,'' or to be intrigued by Sigmund Freud's ''The Erotic Roots of Aviation.'' You don't have to be a flier—but if you are, the pleasure will be all the greater.

Beyond the obvious constraint of book length, I imposed upon myself the further restriction that no writer could be represented more than once, a yoke against which I strained mightily. How can one choose but a single item from the works of Richard Bach or Ernest K. Gann or Antoine de Saint-Exupéry? Making a virtue of necessity, I prefer to think I have introduced some pieces that might not be as well-known in the field of aviation—for example, T. H. White's ''Solo,'' Keven Brownlow's description of the making of the classic film *Wings*, or an excerpt from John McPhee's *The Deltoid Pumpkin Seed*.

I wish to thank my friends and collaborators, David Reuther and Bob Carroll; editors Maron Waxman and Megan Schembre of Scribners; the helpful people at the public libraries of New York and Saugerties; and the particularly gracious Margery Peters of the Library of the General Society of Mechanics and Tradesmen, a splendid source for early aviation material. And for all sorts of in-flight assistance, my wife and copilot, Sharon.

JOHN THORN
Saugerties, New York

Until Abruzzo, Anderson, and Newman crossed the Atlantic in a balloon in August 1978, the only men to have done so were the fictional protagonists of Edgar Allan Poe's "news" story of 1844, published later in this volume. Thirteen actual attempts, the first in 1873, the last only a month before the flight of *Double Eagle II*, had ended in failure. The most recent attempt had been the British *Zanussi*, which fell short of the French coast by a heartbreaking 110 miles; another had been the original *Double Eagle* in 1977 — fifty years after Lindbergh, the Lone Eagle, had made his mark on history.

BEN L. ABRUZZO with MAXIE L. ANDERSON and LARRY NEWMAN

Double Eagle II Has Landed!

Late. Again we were lifting off late, this time by two hours, forty-five minutes.

The year before, our two-man balloon, *Double Eagle*, had been delayed three hours past scheduled launch time, and Maxie and I found a storm that should have been behind us ahead and waiting. Largely because of that delay, our attempt to be the first to cross the Atlantic Ocean by balloon had ended with a frigid and almost fatal ditching in the sea off the coast of Iceland.

Now, at 8:43 P.M. in Presque Isle, Maine, on Friday, August 11, 1978, *Double Eagle II* and her crew of three eased into an almost windless sky above potato fields. Gradually we moved off, people running below us in the last light. An eddy of wind forced us downward. Larry's hang glider, tethered beneath the gondola, nicked the ground and bounced up. I quickly spilled sand ballast, and we climbed into the dark.

It was a comfort to know that, unlike the year before, we left our families and friends dry eyed and cheerful. They had confidence in us, as we had confidence in ourselves, in our craft, and in our ground team.

What was different this time? One difference was the addition of Larry, an extra hand sharing duties and watches to reduce fatigue. He and his associates had rebuilt our catamaran-shaped gondola, utilizing hang-glider parts to make it more efficient and comfortable.

A second difference was experience, both Maxie's and mine as balloon pilots and that of our superb ground team, now on its second time around with us. Maxie puts it this way: "I don't think that you can fly the Atlantic without experience, and that's one reason it hadn't been flown before. Success in any venture is just the intelligent application of failure."

That first night, as we ghosted across Maine and New Brunswick, we relaxed a bit to watch northern lights draping the sky in great shimmering waves as meteor showers blazed across our view.

As we sailed on, a failure from the first attempt emerged again despite all our application to avoid it—the radios worked poorly or not at all.

Our communications director, Sid Parks, had put long hours into making sure the radios worked properly, and on the ground they did. But as Larry says, "Things go up in aircraft and they quit. It's like magic."

There we were again, wishing for a long string and two tin cans. We were flying by the seat of our pants, but at least this time the weather was good.

Our announced goal was not just a crossing from continent to continent, but from the United States to Paris. To plan the flight, we had talked endlessly with Bob Rice, chief meteorologist at Weather Services Corp. of Bedford, Massachusetts, and with our in-flight director, physicist Rich Schwoebel, whom Maxie calls "the most brilliant man I know." Together we plotted strategy and tactics unique to long-distance ballooning and attempts on the Atlantic.

Unlike other balloonists who had tried it, we planned to climb to high altitude quickly and maintain it. As it turned out, for about a third of the flight we were above 15,000 feet and breathing oxygen. At these higher altitudes we could expect stronger winds to speed us along.

Yet going to high altitude is, by itself, not enough. If you just drift with the winds, you can end up going anywhere. Our plan was to pilot the *Double Eagle II* by moving into a migratory high-pressure system of a kind fairly common in the North Atlantic between the storms of spring and autumn.

Such a high-pressure system is basically a mass of air that rotates slowly clockwise as the whole mass moves from west to east. By launching when we did, and despite the delay, we climbed into the heart of the high, which was now squeezed into a ridge by two low-pressure systems, one ahead of us and one behind. If the high kept its shape and strength, we could ride to Europe with it. And, as it rotated like a giant, sluggish merry-go-round, we would follow a curving path around it and grab the brass ring at Paris.

If all went well, that is. If all went badly, W. C. "Doc" Wiley was again ready to coordinate air-sea rescue operations.

At the beginning things went neither way. Our course was good, but our altitude was lower than we wanted. After about forty hours aloft we abruptly lost 3,500 feet in a downdraft. Maxie and I jettisoned ballast calmy and carefully—our experiences from the first flight paid off here—and we began to rise. To make sure we weren't leaking helium, I climbed on Maxie's shoulders, opened the appendix at the bottom of the gas envelope, and scanned the inside.

Even the smallest leak would admit sunlight and appear as does a tiny star in a black sky. I could not see the faintest pinpoint of light, but I didn't expect to. I knew that Ed Yost, who built our balloon—its neoprene-coated panels cemented with superstrong adhesive—had used the finest materials and the highest standards of workmanship.

Indirectly Ed was responsible for our trying this in the first place. Maxie and I were both hot-air-balloon enthusiasts, but neither of us had considered the Atlantic crossing in a helium balloon. Not until one evening.

"I was in my den," Maxie recalls, "and I picked up the February 1977 *National Geographic* that described Ed's solo attempt on the Atlantic. The article captured my interest, and I thought it over for about a week.

"Then I brought it up to Ben—he's the luckiest man I know. He thought it over for a minute. Maybe two; he's not known for being impetuous. He said, 'Let's do it.' From that was born our first flight."

Maxie and I went to see Ed, a man of few words and those well chosen. He agreed to build the *Double Eagle*, and later he built *Double Eagle II*. So here we were, climbing to altitude over open ocean on our second try in less than a year.

By dawn of the second full day out, our flight profile looked good—altitude correct, position in the high-pressure system correct, directly over Newfoundland. Early in the afternoon we passed St. John's, right over the spot where Marconi had received the first transatlantic wireless signal, beamed from Cornwall, England. We could have used him. Our radios continued to cause problems for communications, for navigation, and for Larry.

"It was really a trying time for me," he says, "because both Ben and Maxie were upset about the radios, and I just seemed to be there playing with them. They didn't like it, and they told me so.

"On the third day out, I disconnected all the cables except the ones to the amateur-band radio, and we got our first transmission from a man in England with the call letters G4JY. He came in so loud and clear it almost knocked me out of the gondola. With his help we set up a network all the way back to the ground team in Bedford, northwest of Boston.

"For morale, it was remarkable. It was like a child finding his mother in a crowd."

Even with everything going well, we had little time to sightsee or to relax. Communications, navigation, ballast planning and execution, updating logs, and housekeeping took up 90 percent of our waking hours. Oxygen maskes rasped over our growing stubble. Our thirst increased, but our appetites waned. Maxie disciplined us to eat, and we had everything from soup to salami to nuts, ordinary foods of the kind we ate at home, and plenty of juices.

One of the many hard-knock lessons we learned on the first attempt was that goose-down garments are worthless when wet. This time we dressed in layers, using plenty of wool, as well as synthetics and down. I even had battery-powered electric socks for the coldest nights, since one of my feet had been severely frostbitten on that first voyage. The foot had throbbed with intermittent but excruciating pain for months, a sensation I did not want repeated.

Although we had hoped for better speed than the approximate 15 knots we were averaging, our track was just about as planned. But as Bob Rice says, "The weather is what it is; later we figure out what's average."

During the darkness following our fourth sunset aloft, we got a chill reminder of our ill-fated attempt. Again, as they had near Iceland the year before, nodules of ice began to form on the balloon. Then, at 5,500 feet, an accumulation of 900 pounds had almost weighed us into the sea. This time, at 16,500 feet, the ice was less, about 300 pounds, and after the sun came up it began to melt, falling around the gondola first as ice crystals, then as ice flakes, and, finally, as our own private rain shower.

The North Atlantic was not done with us yet. We knew a low-pressure storm had been dogging along behind us, and Maxie was the first to see it, just hours after the ice was gone.

"When I came on watch I looked to the west, and the storm was there within sixty miles of us. The leading edge pointed fingers of high clouds right at *Double Eagle II*, and I looked at it for quite a while.

"By this time we were already climbing on our way to 23,500 feet, and the chances were good that the storm would veer to the north, which it did—and it turned into a full-blown dandy."

As the storm turned, so did we but in another direction, leaving our northeast course as we climbed through the high-pressure ridge and curved southeastward away from Iceland toward the British Isles. Our greatest danger was yet to come, and it was upon us as soon as the storm left our track.

To understand the predicament we faced, it is necessary to understand something of how a helium balloon is piloted. Lighter-than-air helium in the gas envelope makes the balloon rise, whereas ballast attached to the gondola or carried inside as equipment prevents the balloon from rising too far or too fast. Ballast is the "fuel" of a balloon, since releasing some of it overboard is the only way to stop a descent.

So far, so good, but there is a complication. As a result of releasing ballast on previous days, our balloon now weighed less.

Moreover, in daytime, as the sun reheats the helium that has contracted in cool night temperatures, the gas expands again to fill the envelope to maximum volume and lifting capacity. Day by day the balloon wants to go higher and higher, and you can get caught in a yo-yo effect, expending excessive amounts of both helium and ballast—burning your candle at both ends—just to stay aloft.

Our strategy, much easier said than done, was to conserve both by maintaining near equilibrium at an ever-increasing altitude.

After our back-to-back episodes with the ice and the storm, the sky above cleared to a deep blue. But by noon three factors conspired against us. The sun stood overhead, beating down only on the top of the gas envelope, sprayed with aluminum paint to reflect heat. Then some high cirrus clouds moved in and screened the sun, a condition we call "partial sunset." Cooled, the balloon began to descend from 23,500 feet.

But still another factor was at work, one we would theorize about later. Whatever it was, it was pushing or sucking us right down into the very center of a perfect circle punched in the cloud deck below.

We faced a difficult choice. If we ballasted off heavily and at once, we would pop up like a beach ball released underwater. We would not only waste precious ballast but also would soar through our weight-determined ceiling of 26,500 feet.

If we ascended to 29,000 feet, the gas envelope would automatically vent 26,000 cubic feet of helium. In that event we would then have to drop an additional 800 pounds of ballast at night to stay aloft. Result: We would be short on ballast, and unless we were to saw off the bow and stern of the gondola for ballast, we would fall into the sea short of Ireland.

Instead of throwing ballast over in panic, we ballasted carefully, slowing our rate of descent from 400 feet per minute to about 200 feet.

We continued to drop from our high of 23,500 feet, past 20,000, into the teens, past 10,000. And we kept descending. As we neared 4,000 feet, Larry yelled, "Ben, we're in the clouds."

Larry is an experienced hang-glider and airplane pilot—he soloed at age twelve—but this was only his second time in a balloon. No wonder he was anxious. As Larry recalls: "I was really upset, especially since we had been doing so well and seemed so close to success. I said to Ben and Maxie, 'You know, guys, I don't think we're going back up.'

"'Yeah, we are,' they said.

"I said, 'No, we're not. Tell you what. I'll bet you a hundred dollars we won't go back above 12,500.

"'Make it ten dollars,' said Maxie.

"'Ten dollars—you're crazy! Let's make it a hundred.'

"Sure enough, the balloon finally bottomed out, thanks to the ballast going over, and about thirty minutes later we passed 12,500 feet. I couldn't have been happier. I just threw the hundred dollars at them."

Superheating from the afternoon sun drove us up to 24,950 feet, where, with the cool of dusk, the balloon stabilized. We had come through the big drop without wasting helium or ballast—we had dumped just 300 pounds.

We can only make an educated guess as to what had happened. Here it is: We had been stable in a cool, buoyant air mass. Then the balloon floated into a mass of descending dry air, which grew warmer as it compressed. When that happened, the balloon was no longer in equilibrium, and it dropped down that falling shaft of warmer, less buoyant air.

We had been expending ballast wisely, but as we neared land, not cheaply. We couldn't jettison heavy items over populated areas, so, as planned, over went such gear as tape recorder, radios, film magazines, sleeping box, lawn chairs, most of our water, food and the cooler it was in, and our spent oxygen tanks.

At sundown off Ireland, it also came time to ditch the hang glider Larry had hoped to fly from the gondola to a landing in Europe.

"I knew it had to go," Larry says. "When I cut it loose, it dropped below us, then came back up only five feet away. It dropped again, did three tight, perfect loops and a series of turns. It took almost thirty minutes to disappear into the clouds below."

As night edged out from Europe to engulf us, the full moon shone brightly on the cloud deck below and a line of towering cumulus that was rising westward. It was at such moments that I wished my family, all my friends—everybody really—could be in the gondola with me.

Not long after that lovely, silent pause, we received a transmission from air control at Shannon: "*Double Eagle II*, you are over the coast of Ireland."

Larry replied, "Shannon, this is *Double Eagle II*. Are you sure?"

"Yes, sir. Our radar is never wrong."

Strangely, perhaps, we felt no great elation. We were crushed by fatigue and, with our oxygen thrown over, plagued by headaches. Paris, our goal, was still a long, uncertain way off, though Maxie had another definition of success.

"If it's dry below, it counts."

We began to outpace the clouds, and Dublin came brilliantly into view to the northeast. Below, through the scattering clouds, we picked out the occasional wink of lights from the ground as we ballasted off our excess clothing.

We unpacked our survival suits, much like wet suits, unzipped them, and threw them over. As they fluttered down, they looked like human shapes swirling above the clouds. I can only imagine someone on the ground watching them swoop downward, and I wonder if tales of banshees were told that night.

As light began faintly to break in the east, we departed Ireland, crossed St. George's Channel, and made landfall above Strumble on the Welsh coast. The landscape unrolled beneath us, green and glowing, a magnificent sight after days of nothing but blue sea and sky and white clouds.

Aircraft now swarmed around us, some too close for comfort. A Royal Air Force jet fighter snapped a series of rolls in salute. The plane we were happiest to see circled with our wives aboard. It was their day as much as ours.

Maxie says that after we made Ireland he "felt like a tourist." Well, maybe, but we had serious flying ahead. Paris was our announced goal, and Paris we would try to make. Doc Wiley stood by in London, using available channels of communication to direct us.

During the cool shank of the night we had dropped to 11,000 feet. Now, with the sun superheating the helium, we climbed to 15,000 feet over southern England. Had we let it, the balloon would have soared higher and higher—up to 29,000 feet before losing helium automatically and stabilizing. Word came from Rich Schwoebel, "You're at 15,000. Descend to 10,000. Course good for Paris." The message confirmed a decision we had already made, based on winds-aloft information we received from a weather station in England.

To descend deliberately within the desired vector, I had to valve off helium through an aperture at the very top of the balloon. I opened it by means of a long lanyard. If we stayed too high, the winds would bend our course eastward over London for a probable landing in Belgium. If we descended too quickly, we would curve south toward Brittany.

We had *flown* across the Atlantic, maneuvering to take advantage of favorable winds, not just drifting. Before us was the final test of our finesse as balloon pilots.

As we passed over England, we were too high to see people on the ground. They were there, no doubt about it. From every village and town, farm and roadside, the flash of mirrors sparkled up at us. Everybody was shining mirrors, it seemed, everybody. So were we. We signaled back down to the lush countryside sequined with flashes.

Approaching the English Channel at Bournemouth, we were apprehensive. Our course across that notoriously rough stretch of water was a diagonal some 120 miles long. Larry remembers it well.

"All the numbers were bad. The track was about the longest way you could go. It was the worst time of day, near noon, the same time we had our big sink the day before. We had no oxygen and little ballast."

Were we pushing our luck? Maxie has considered the part luck played in our voyage.

"A measure of good fortune was required, but success does not depend on it. It may tip the scales, like the final grain of sand. But like success in any undertaking, the requirements are dedication, preparation, and work."

Finally, as the sun was lowering, we crossed the channel, floating over Le Havre, France. That was the big one. Then we were jubilant.

We descended very gradually, between 100 and 200 feet per minute. Just by great coincidence, the scattered cumulus clouds below parted behind us, leaving a clear pathway back to Le Havre as precise as a canal in the air. Ahead, the clouds were dissipating. We would seem headed directly for a cloud. When we got there, it was gone.

Features on the ground became sharper and sharper as we dipped lower and lower.

Over the radio came word that for our sake authorities had closed the busy airfield at Le Bourget, where Lindbergh had landed. We declined the offer; passing over the suburbs of Paris was too risky, both for us and for the citizens.

As we sank ever lower, we started to look for a place to drop our remaining heavy ballast and for a landing site. It would have been duck soup for Larry in his hang glider, but as he says:

"We were running out of places where we could drop ballast or land, when all of a sudden there was the town of Evreux.

"We spotted some plowed fields, descended to 300 feet, and dropped an empty propane tank and a battery. We watched them fall harmlessly and heard the impact: *thump, thump.*"

That gave us about 2,500 feet of altitude to clear Evreux, and we began to look for a landing site near the small village of Miserey.

Below, we could see cars everywhere and wondered, "What's going on here?" Then we realized. They were strung out along our course and converging like the spokes of a wheel. We began to see people on the ground and waved, but not much; we were numb tired. Maxie and I weren't worried about having enough energy for the landing. Adrenaline, we knew, would take over.

At last we were committed to a landing, with one power line to clear and a green, then a golden field to set down in. The long journey was all but ended.

Now, months later, each of us recalls certain moments during the aftermath.

There was my sadness that the voyage was done, a feeling I hadn't anticipated and can't quite explain.

Maxie vividly remembers the mayor of Evreux giving a speech of welcome from a balcony.

"It could have been the mayor of any small city in the United States on the Fourth of July. The only barrier that existed between us and the French was one of language."

Then the celebration swirled on to Paris, and I cannot forget the small woman in the white dress with a pattern of red flowers. She had a little camera that couldn't have cost more than six or seven dollars. Except for that first time, she never asked for an autograph. She showed up everywhere we went, including private functions that no one knew about, not even the press.

There was Larry partying in Paris, while we two—shall I say, more experienced balloonists—collapsed into our beds. And there was Larry feeling what each of us sensed: an absolutely genuine joy, a shared joy, in our adventure.

Frankly, until we landed, I never understood fully what had motivated me, the real reason for trying what had ended in grief for so many before us. People may not be able to picture themselves in a spacecraft to the moon or beyond, but they can see themselves in the balloon. They can feel with us that when mankind stops crossing frontiers or achieving new goals, it stagnates and moves backward instead of forward. That moving forward was, for me, the long-hidden motive of our voyage.

Everywhere, it seemed, there were speeches and celebrations, medals and honors. Albuquerque, the adopted hometown of each of us, turned itself inside out.

But the landing itself was the most powerful memory of all.

In those final moments we approached earth again, dead tired and six days dirty. Just before Larry cut the wires to stow the radio, we made our last transmission.

"All aircraft in the area, *Double Eagle II* is landing."

This boisterously morbid drinking song of World War I is testament to the fighter pilot's iden-
tification with his plane and to his fatalistic view that he is just a replaceable piece of hardware.
Aviators of the Lafayette Escadrille and Flying Corps sang this to the tune of "The Old Oaken
Bucket"; their counterparts in the Royal Flying Corps took as their model "My Old Tarpaulin
Jacket."

ANONYMOUS

A Poor Aviator Lay Dying

British Version

Oh, a poor aviator lay dying
At the end of a bright summer's day,
His comrades were gathered around him
To carry the fragments away.

The engine was piled on his wishbone,
The Hotchkiss was wrapped round his head,
A spark plug stuck out of each elbow,
It was plain that he'd shortly be dead.

He spat out a valve and a gasket
And stirred in the sump where he lay,
And then to his wond'ring comrades
These brave parting words he did say:

"Take the manifold out of my larynx
And the butterfly valve off my neck,

Remove from my kidneys the camrods,
There's a lot of good parts in this wreck.

"Take the piston rings out of my stomach,
And the cylinders out of my brain,
Extract from my liver the crankshaft,
And assemble the engine again.

"I'll be riding a cloud in the morning,
With no rotary before me to cuss,
So shake the lead from your feet and get busy,
There's another lad wanting this 'bus."

Who minds to the dust returning?
Who shrinks from the sable shore
Where the high and the lofty yearning
Of the soul shall be no more?

So stand to your glasses steady.
This world is a world full of lies.
Here's a health to the dead already,
And hurrah for the next man who dies!

American Version

The young aviator lay dying,
And as 'neath the wreckage he lay,
To the mechanics assembled around him,
These last parting words he did say:

"Two valve springs you'll find in my stomach,
Three spark plugs are safe in my lung,
The prop is in splinters inside me,
To my fingers the joy stick has clung.

"Take the cylinders out of my kidneys,
The connecting rods out of my brain;
From the small of my back get the crankshaft,
And assemble the engine again."

We meet 'neath the sounding rafters,
The walls all around us are bare.
They echo the peals of laughter;
It seems that the dead are there.

So stand by your glasses steady,
This world is a world of lies.
Here's a toast to the dead already;
Hurrah for the next man who dies.

Cut off from the land that bore us,
Betrayed by the land that we find,
The good men have gone before us,
And only the dull left behind.

So, stand to your glasses steady,
The world is a web of lies.
Then here's to the dead already;
And hurrah for the next man who dies.

They don't write 'em like this any more, for which one may be wistful or grateful. The inaugural opus in the Tom Swift series, *Tom Swift and His Motorcycle*, appeared in 1910, followed in the same year by the first of the boy inventor's many air adventures. In the book from which the passage below is taken, *Tom Swift Circling the Globe*, our hero and his *Air Monarch*, a unique vehicle that can travel as easily on land or water as in the air, are battling time, circumstance, and most of all the nefarious crew of the rival *Red Arrow*. If Tom can traverse the necessary 25,000 miles in twenty days, he will win a stupendous wager placed by his proud father as well as the approbation of his sweetheart, Mary Nestor. The prolific Victor Appleton, by the way, is a collective pseudonym employed by the mind-bogglingly industrious writers of the Stratemeyer Syndicate, the folks who gave us Don Sturdy, the Rover Boys, Baseball Joe, and countless other idols of a bygone age.

VICTOR APPLETON

Tom Swift in Dire Straits

Tom Swift for a moment was torn between duty and ambition.

His machine was winging along at wonderful speed and he was beginning to make up for much time lost. To slow up, descend and rescue these two on the raft meant more delay—a delay that would be dangerous to his chances of winning the prize. He did not know how many or what other ships, whether of the air or the sea, containing his rivals, might be ahead of him or close behind.

But it was for only a moment that Tom hesitated. He gave one look down at the despairing, helpless men on the raft and cried to Ned: "We'll go down!"

Ned knew, as well as Tom, what this might mean.

As the young inventor sprang into the motor room to give the order to Hartman, who was on duty, he practically gave up all hope of winning the race. Yet he had no regrets.

There was another thought that came to Tom as he told the surprised Hartman what was about to be done and mentioned the raft with the shipwrecked ones on it. This was the problem of caring for the two castaways when they were taken aboard the *Air Monarch*.

"There's hardly room for them," reasoned Tom. "Their added weight will hold me back, even if I'm able to make up any of this lost time. And we haven't any too much food. Didn't have a chance to lay in any at the camps of the pirates and headhunters," he grimly reflected.

But he did not hesitate, and a little later two very thankful, but much wondering, men were being taken aboard the airship. They were thankful for their rescue but surprised at the manner of it.

"We thought some steamer might pick us up," said one, "but we never counted on something coming out of the sky to do it."

"Sam thought I was out of my head when I told him an airship was coming," remarked the other.

Tom had sent his craft slowly over the water on her pontoons as close as he dared to go to the raft, and the men had leaped into the sea, swimming the intervening distance, since it would take but a slight bump from the jagged edges of the raft to puncture the frail body of the *Air Monarch.*

Once on board, and again riding through the air, Tom listened to the stories of the castaways. They were part of the crew of a small lumber schooner that had broken up in a terrible storm. For more than a week the men had been drifting about on the raft which had been made from some of the deckload of lumber. Five of their companions had been washed off, and one, in delirium, had leaped into the sea and was eaten by sharks. The two who were left had only a little food and water remaining when they were saved.

"I'm sorry that I can't take you men all the way back to San Francisco with me," Tom said, when the two had been made comfortable in temporary bunks and given some extra garments in place of their wet and storm-torn ones. "But I'm trying to win a race. How would it do if I landed you on one of the Hawaiian Islands? I've got to stop there for oil and gas."

"That would suit us fine, Captain," said Sam Stout, while his companion, Frank Madler, said: "We can easily get another ship there."

So it was arranged, and Tom, still with a faint hope in his heart that he might at least come in a good second if not the winner of the world race, turned on a little more power and headed for the east. There lay the United States, and once over that territory there remained only the last part of the flight—across the continent.

The motors of the *Air Monarch* were not behaving as well as Tom liked, and he had an idea it was due to the poor quality of the last gasoline he had put into his tanks. He dared not use the last of his super-fuel, but he hoped in Hawaii to get some better than the last.

If worse came to worst, he thought he could finish the race in his *Airline Express* craft, but he wanted to do it in the *Air Monarch.* It would be much more satisfactory, he told Ned, who agreed with him.

It was only half a day's travel from where the shipwrecked ones had been picked up to the harbor of Honolulu, and it was about mid-afternoon when Ned, who was on watch, gave the cry: "Land ho! All out for Hawaii!"

The beautiful islands were looming ahead of them through the mist. Quarter of an hour later they made out Diamond Head and knew they were close to Honolulu, the chief city of the territory.

Tom was in the pilothouse, prepared to make a landing, if such a term is permissible when one means to drop into the water. He had headed the craft for a spot somewhat outside a harbor, intending to taxi up into it to avoid the shipping when, suddenly, Sam, one of the shipwerecked sailors, who was looking from the pilothouse window, pointed to a spot directly in front of them and cried: "There she blows!"

"What?" asked Tom, though a second later he realized what was meant.

"A whale!" cried the sailor. "There she blows, and you're going to bump right into her!"

Tom tried desperately to shift the wheel and, at the same time to elevate the airship to pass over the monster of the deep. But they were now so close that it seemed impossible. With the motors shut off the sound of the whale's blowing could be heard and each moment the vast bulk became plainer. If the airship hit that mountain of flesh she would be instantly wrecked!

"Start the engine! Give me some speed!" Tom yelled desperately. "I've got to zoom!"

He meant, by this, a sudden and sharp lifting of the airship over the whale, as a birdman often zooms to avoid crashing into trees or some obstruction.

Luckily, Peltok was on duty in the engine compartment. He had shut off power but a short time before, and the cylinders were still hot. In a second the machinist switched on the spark, hoping to start the motors on compression as can sometimes be done. To his delight it happened this time.

With a roar the powerful engines started up, whirring the propellers and giving the craft enough momentum for Tom to lift her over the whale's back.

But so little room was there to spare that afterward, observers in nearby boats declared that the spouting of the whale wet the lower portion of the *Monarch.*

Tom could well believe this, for when the big creature, alarmed by the near approach of the aircraft, raised its flukes and slammed them down on the surface of the sea, preparatory to sounding, the water was washed in a big wave over the rudders of the *Air Monarch,* tearing loose some of the stays and guy wires of the elevating surfaces.

It was a narrow escape, and Tom realized this as, a little farther on, he brought his craft safely to the calm surface of the bay while behind him the waves were ruffled by the sinking of the whale that was soon lost to sight.

"If this keeps up," remarked Ned whimsically, as he sat on a locker, "I'll be a nervous wreck after this race. It's just one bit of excitement and narrow squeak after another."

"We have had a little more than our share," admitted Tom. "But I think the worst is over now."

"You sure handled your ship like a veteran!" commended the two shipwrecked sailors.

Tom's arrival at Honolulu was greeted with a great demonstration on the part of officials and the populace, some of whom had expected that one or more of the world racers might pass over their islands. So when word came that Tom had stopped to take on gasoline and oil, arrangements were made to fête him. But he had little time for any ceremonies although he did consent to be decked with a wreath of flowers—a native custom.

"I want to hop off again as soon as I can," he told the welcoming delegation, though as politely as possible. "You understand how it is."

"Oh, yes, we understand," was the reply. "But one of your rivals is here, and he seems to be taking his time."

"Who is it?" asked Tom, though he was almost prepared for the answer that came.

"Dan Kilborn in the *Red Arrow*."

"Here ahead of us!" exclaimed Ned.

"That isn't to be wondered at!" remarked Tom. "The thing for us to do is to leave ahead of him and keep him at a distance."

They learned that the *Red Arrow* had arrived two days before with a broken cam-shaft and that the repairs were nearly completed. On hearing this Tom hastened as much as he could the taking on of gas, oil, and other necessities. But when it seemed that they might get under way again a few hours after landing in Honolulu, Peltok discovered another small burned-out bearing that must be replaced.

"It will not take long," he said, "as we have spare parts for that. By night we can be moving again."

"I hope so," murmured Tom.

The two shipwrecked sailors were taken in charge by the captain of a vessel who promised them berths, and Tom and Ned sent home radiograms telling of their progress up to date.

In spite of Peltok's assertion that it would not take long to replace the burned-out bearing, it did, and he had to amend his calculation so that it would be midnight before the *Air Monarch* could take off again.

Tom and Ned occupied their time by visiting places of interest, and it was when they were coming out of a restaurant that they saw a crowd approaching them. Thinking it was only curious ones who wanted to look at the "world fliers," the two young men paid little heed until they heard a voice they knew saying:

"There's Tom Swift now! Arrest him! I'll make the charge!"

Tom and Ned wheeled about to see Dan Kilborn facing them. The pilot of the *Red Arrow* was in company with a police officer, and again he exclaimed: "Arrest Tom Swift!"

"On what charge?" asked the officer.

"He tried to kill me!"

"Kill you!" shouted Tom. "Are you crazy?"

"No, I'm perfectly rational!" sneered Kilborn. "But I make that charge. A charge of attempting my life! Tom Swift dropped from his airship a Chinaman on my head, severly injuring me."

And then it came to Tom and Ned what the rascal meant. He was referring to the time he had set the Chinese bandits on to wreck the *Air Monarch*. One of the bandits had been carried up by catching hold of a rope as Tom sent his craft aloft, but the frightened fellow had loosed his hold and dropped on Kilborn's head.

"Arrest Tom Swift!" again demanded the *Red Arrow* pilot.

As he hastened forward, so did the police officer, accompanied by a number of others.

"I am sorry," said the officer to Tom, "that I shall have to take you into custody. There must be a hearing, but probably, since no one was really killed, you will be admitted to bail."

"You mean that I must submit to arrest and proably lose a day, if not more, arranging for bail on this untrue charge?" asked Tom indignantly.

"Such is the law," was the answer.

"It's a foolish law!" cried Ned. " It was Kilborn's own fault that the Chinese bandit dropped on him. He sent them to attack us!"

"I did nothing of the sort!" declared Kilborn brazenly.

"I must take you into custody, young man," said the officer. "I am sorry, but this gentleman," and he pointed to Kilborn, "has sworn out a warrant against you, charging you with assault with intent to kill. I must do my duty."

"All right," assented Tom, with such seeming cheerfulness that Ned looked at him curiously. "If I have to go with you I suppose I must. But this is your last trick, Kilborn!" the young inventor suddenly cried. "I'm going to play trumps from now on! Follow me, Ned!"

With a sudden motion Tom tripped the officer who had reached out a hand to apprehend him. He pushed the man backward into the midst of his fellows, and then sent a fist full into Kilborn's face, whirling him aside.

Then, like a football player, Tom turned and ran back into the restaurant, followed by Ned, who did not know what to make of it.

"They'll trap us in here, Tom!" panted his chum.

"No, there's a back way out that leads directly to the beach!" whispered Tom. "I noticed that when we were in there. Come on. We can beat Kilborn yet!"

On they rushed, through the midst of the astonished waiters and patrons in the dining room. Out through the kitchen they went and into a back alley. Tom had marked the way well, and in a few minutes, leaving a confused and yelling crowd of men behind them, the two reached the harbor, and, engaging a motor launch by the simple but effective method of shoving gold coin into the owner's hand, were soon aboard the *Air Monarch*.

"How about it?" gasped Tom to the workmen. "Can we start?"

"At once, if there is need!" answered Peltok.

"There's the greatest need in the world if I'm going to win the race!" cried Tom.

A minute later the *Air Monarch* rose.

Legend has it that the first men to fly were Daedalus and the ill-fated Icarus. History gives that honor to Jean-François Pilâtre de Rosier and François Laurent, the Marquis d'Arlandes. They ascended into the heavens only two hundred years ago, in a hot-air balloon designed by the brothers Montgolfier. The flight had been preceded by an unmanned test on June 4, 1783, and by another on September 19, in which a duck, a rooster, and a sheep were sent aloft. The latter event had been witnessed by Louis XVI, who was much impressed but, upon learning that the Montgolfiers intended next to send up a man, insisted that only a criminal be risked in so perilous a venture. Pilâtre de Rozier, a daredevil of sorts, thought that the glory of the ascent should not fall on one so base and enlisted the aid of the Marquis d'Arlandes to convince the king. Louis gave his assent, and the marquis insisted that, as reward for his powers of persuasion, he be allowed to come along as passenger.

MARQUIS D'ARLANDES

Man's First Flight

We went up on the 21st of November, 1783, at near two oclock, M. Rozier on the west side of the balloon, I on the east. The wind was nearly northwest. The machine, say the public, rose with majesty. . . .

I was surprised at the silence and absence of movement which our departure caused among the spectators, and believed them to be astonished and perhaps awed at the strange spectacle. . . . I was still gazing, when M. Rozier cried to me—

"You are doing nothing, and the balloon is scarcely rising a fathom."

"Pardon me," I answered, as I placed a bundle of straw upon the fire and slightly stirred it. Then I turned quickly, but already we had passed out of sight of La Muette. Astonished, I glanced at the river.

"If you look at the river in that fashion you will be likely to bathe in it soon," cried Rozier. "Some fire, my dear friend, some fire!"

We traveled on . . . dodging about the river, but not crossing it.

"That river is very difficult to cross," I remarked to my companion.

"So it seems," he answered; "but you are doing nothing. I suppose it is because you are braver than I, and don't fear a tumble."

I stirred the fire, I seized a truss of straw with my fork; I raised it and threw it in the midst of the flames. An instant afterwards I felt myself lifted as it were into the heavens. . . .

MAN'S FIRST FLIGHT,

IN AN UNTETHERED BALLOON AT PARIS ON NOVEMBER 21, 1783, WAS NEARLY MADE BY A CONVICTED MURDERER.

KING LOUIS XVI, WHO'D WATCHED AN EARLIER, UNMANNED FLIGHT, PLANNED TO PROVIDE A CONVICT FOR THE EXPERIMENT.

BUT...

...THE KING'S HISTORIAN, FRANÇOIS PILATRE de ROZIER, CONVINCED THE KING THAT THE GLORY SHOULDN'T GO TO A FELON AND OFFERED HIMSELF AS PILOT.

THE MARQUIS d'ARLANDES, WHO COURAGEOUSLY SERVED AS THE FIRST AIR PASSENGER, WAS LATER CASHIERED FOR COWARDICE DURING THE FRENCH REVOLUTION.

At the same instant I heard from the top of the balloon a sound which made me believe that it had burst. . . . As my eyes were fixed on the top of the machine I experienced a shock, and it was the only one I had yet felt. The direction of the movement was from above downwards. . . .

I now heard another report in the machine, which I believed was produced by the cracking of a cord. This new intimation made me carefully examine the inside of our habitation. I saw that the part that was turned toward the south was full of holes [from sparks], of which some were of a considerable size. . . .

I took my sponge and quietly extinguished the little fire that was burning some of the holes within my reach; but at the same moment I perceived that the bottom of the cloth was coming away from the circle which surrounded it. . . .

I then tried with my sponge the ropes which were within my reach. All of them held firm. Only two of the cords had broken. . . .

We were now close to the ground, between two mills. As soon as we came near the earth I raised myself over the gallery, and leaning there with my two hands, I felt the balloon pressing softly against my head. I pushed it back, and leaped down to the ground. Looking round and expecting to see the balloon still distended, I was astonished to find it quite empty and flattened. . . . I saw [Rozier] in his shirt-sleeves creeping out from under the mass of canvas that had fallen over him. . . . we were at last all right.

Poetry or prose? Romanticism or classicism? Yin or yang? The question is not simply how one views one's relationship with a plane and with the sky. The old-timers will tell you that flying went out with doped linen and singing wires. They are wrong. The spirit of flying is not in the airplane but in the airman, and no one knows that as well, or expresses it as well, as Richard Bach.

RICHARD BACH

Aviation or Flying?
Take Your Pick

You look at aviation and you can't help wondering. There is so much going on all at once, and the whole thing is so foreign and complicated, and there are so many roaring individualists there, all railing at each other over tiny differences of opinion.

Why would anyone, you ask, deliberately dive into that maelstrom, just to become an airplane pilot?

At the question, the tumult stops instantly. In the dead silence, the pilots stare at you for not knowing the clearly obvious.

"Why, flying saves time, that's why," says the business pilot, at last.

"Because it's fun, and no other reason matters," says the sport pilot.

"Dummies!" says the professional pilot. "Everybody knows that this is the best way in the world to make a living!"

Then the others are at it again, all talking at once, and then shouting for your attention.

"Cargo to haul!"

"Crops to spray!"

"Places to go!"
"People to carry!"
"Deals to close!"
"Sights to see!"
"Appointments to keep!"
"Races to win!"
"Things to learn!"

They are at each other's throats once more, snarling over which part of the gold of flight gleams more brightly than any other. You can only shrug your shoulders, walk sadly off, and say, "What could I expect? They're all out of their minds."

You speak more truly than you might think. The government of pure reason departs when an airplane enters the scene. It is no secret knowledge, for instance, that a tremendous number of business airplanes are purchased because someone in the company likes airplanes and wants one around. Given the desire, it is a simple matter to justify the company's ownership of the airplane, because an airplane is also a very useful, time-saving, moneymaking business tool. But the desire came first, and then, later, the reasons were trotted out.

On the other hand, there are still some company executives whose fear of airplanes is as irrational as the affection of others, and despite time or money, saved or earned, have it clearly understood that their company will positively have nothing to do with any flying machines.

For a great many people around the world, an airplane has a special charm that time cannot dissolve, and a simple test illustrates the point. How many things are there on earth today, dear reader, that you truly and deeply want to own, with that same intense longing-to-possess that you had for that metallic blue Harley-Davidson when you just turned sixteen?

So often, as we grow, we lose the capacity to want things. Most pilots are absolutely uncaring about the kind of automobile they drive, the precise form of the house they live in, or the shape and color of the world about them. Whether or not they have or don't have any particular material thing is not of earthshaking importance. Yet it is common to hear those very men openly hungering after one specific airplane, and to see them making huge sacrifices for it.

Rationally speaking, most pilots can't afford to own the airplanes that they do. They give up a second car, a new house, golf, bowling, and three years' lunch just to keep that Cessna 140 or a used Piper Comanche waiting for them in the hangar. They want these airplanes, and they want them almost desperately. More than the Harley-Davidson.

The world of flight is a world in its youth, that is ruled by emotion and hard impulsive attachments to airplanes and ideas about airplanes. It is a world that has so many things to see and do that it hasn't had time for mature reflection about itself, and because of this, like any youth, it is none too sure of its own meaning or reason for its existence.

There is a tremendous difference, for instance, between "Aviation" and "Flying," a difference so vast that they are virtually two separate worlds, with precious little of anything in common.

Aviation, far and away the largest of the two, comprises the airplanes and airmen who have interests beyond themselves. Aviation's big advantage is the obvious one: airplanes can compress a very large distance into a very small one. If New York is just across the street from Miami, one might cross that street three or four times a week, just for the change of scenery and climate. The Aviation enthusiasts find that not only is New York just across the street, but so are Montreal, Phoenix, New Orleans, Fairbanks, and La Paz.

They find that after a very modest amount of training in the not-too-difficult mechanics of the airplane and the not-too-complicated element of the air, they can constantly feed their insatiable appetite for new sights, new sounds, for new things happening that have never happened before. Aviation offers Atlanta today, St. Thomas tomorrow, Sun Valley the next day, and Disneyland the next. In Aviation, an airplane is a clever swift traveling device that lets you have lunch in Des Moines and supper in Las Vegas. The whole planet is nothing but a great feast of delicious places for the Aviation enthusiast, and every day for as long as he lives he can savor another delicate new flavor of it.

To the Aviator, then, the faster and more comfortable his airplane, and the simpler it is to fly, the better suited it is to his use. The sky is the same sky everywhere, and it is simply the medium through which the Aviator moves to reach his destination. The sky is nothing more than a street, and no one pays any attention to the street, as long as it leads to far Xanadu.

The Flyer, however, is a different creature entirely from the Aviator. The man who is concerned with Flying isn't concerned with distant places off over the horizon, but with the sky itself; not with shrinking distance into an hour's airplane travel, but with the incredible machine that is the airplane itself. He moves not through distance, but through the ranges of satisfaction that come from hauling himself up into the air with complete and utter control; from knowing himself and knowing his airplane so well that he can come somewhere close to touching, in his own special and solitary way, that thing that is called perfection.

Aviation, with its airways and electronic navigation stations and humming autopilots, is a science. Flying, with its chugging biplanes and swift racers, with its aerobatics and its soaring, is an art. The Flyer, whose habitat is most often the cockpit of a tailwheel airplane, is concerned with slips and spins and forced landings from low altitude. He knows how to fly his airplane with the throttle and the cabin doors; he knows what happens when he stalls out of a skid. Every landing is a spot landing for him, and he growls if he does not touch down smoothly three-point, with his tailwheel puffing a little cloud of lime-dust from his target on the grass.

Flying prevails whenever a man and his airplane are put to a test of maximum performance. The sailplane on its thermal, trying to stay in the air longer than any other sailplane, using every particle of rising air to its best advantage, is Flying. The big war-surplus Mustangs and Bearcats, moaning four hundred miles per hour down their racing straightaways and brushing the checkered-canvas pylons on the turns, are Flying. That lonely little biplane way up high in a distant summer afternoon, practicing barrel rolls over and over and over again, is Flying. Flying, once again, is overcoming not the distance from here to Nantucket, but the distance from here to perfection.

Although he is in a very small minority, the Flyer is allowed to walk both his own world and the world of Aviation. Any Flyer can step into the cabin of any airplane and fly it anywhere that an Aviator can. He can overcome distance any time it strikes his fancy.

An Aviator, however, isn't capable of strapping himself into the cockpit of a sailplane or a racer or an aerobatic biplane and flying it well, or even flying it at all. The only way that he can do this is to enter the same long training that ironically transforms him into a Flyer by the time he has gained the skill to operate such airplanes.

Far from the relatively simple process of learning to aviate, Flying rears itself a gigantic towering mountain of unknowns to the fledgling, so that where Flyers are, one often hears the cry, "Good grief, I can never learn it all!" And of course it's true. The professional aerobatic pilot, or air racer or soaring pilot, practicing every day for years, is never caught saying, even to himself, "I know it all." If he stops flying for three days, he can feel the rust when he flies on the fourth. When he lands from his very best performance, he knows that he still has room to improve.

Bring these two worlds together in any but the same man, and sparks fly. To the distance-conquering Aviator, the Flyer is a symbol of irresponsibility, a grease-stained throwback to the days of flight before Aviation came to be; the very last person one would exhibit to the general public if one would wish Aviation to grow.

To the skill-seeking Flyer, the unskilled world of Aviation has already grown too much. The poor Aviatiors, he says, don't really know their airplanes when they are performing any maneuver but level flight, and they are the ones who, not caring to study their machines or the face of the sky, turn themselves daily into stall-spin statistics. They are the ones who press on into bad weather, not knowing that without the ability to fly on instruments, those clouds are just as deadly to them as pure methane gas.

"No one is so blind as the man who refused to see," the Flyer quotes in ill-concealed distaste over any pilot who does not share his own zeal to know and to completely control any airplane he touches.

The Aviator believes that air safety is the result of proper legislation and strict enforcement of the rules. The Flyer believes that perfect safety in the air means the ability of a pilot to perfectly control his airplane; that any airplane, perfectly controlled, will never have any accident unless the pilot wishes to have one and controls the airplane into it.

The Aviator tries his level best to obey every regulation he knows. The Flyer is often airborne when regulations forbid it, yet just as often refuses to fly under other conditions that are quite legal.

The Aviator trusts that the modern engine is very well designed and will never stop running. The Flyer is convinced that any engine can fail, and he is always within gliding distances from some suitable place to land.

It is the same sky over both, the same principle keeps both men and both machines aloft, yet the two attitudes are so different as to be farther apart than miles can measure.

So the newcomer, from his very first hour in the air, is faced with a choice that must be made, though he may be unaware he is making any choice at all. Each world has its own special joys and its own special dangers. And each has its own special kind of friendships formed, that are an important part of any life above the earth.

"Well, we defied gravity one more time." Reflected in that common after-flight saying is a hint of the tie that binds airmen together, each in his own world. Airborne, the airman is matching himself against whatever the sky has to offer. The sky and the airplane combine in a challenge, and the airman, Aviator or Flyer, has decided to accept that challenge. The far-traveling Aviator has friends of similar thought and decision all over the country; his circle of friends has a radius of a thousand miles. His counterpart, the Flyer, makes his own fierce friendships, bound as he is in a defensive minority which is convinced of the rightness of its principles.

Why fly? Ask the Aviator and he will tell you of faraway lands brought right to where you can see and touch and hear and smell and taste them. He will tell you of crystal blue seas waiting in Nassau, of the bright clattering casinos and the smooth quiet river at Reno, of the horizon-wide carpet of solid light that is Los Angeles after dark, of marlin leaping up from the ocean at Acapulco, of history-soaked villages in New England, of blazing desert sunsets as you fly down through Guadalupe Pass into El Paso, of Grand Canyon and Meteor Crater and Niagara and Grand Coulee from the air. He will urge you into his airplane, and in moments you'll be covering two hundred miles per hour to some favorite place with a magnificent view and where the chef is his special friend. Back at the airport after a night flight home, locking his airplane, he'll say, "Aviation is worth your while. More than worth your while. There is nothing like it."

Why fly? Ask the Flyer and he will pound on your door at 6 A.M. and whisk you to the airstrip and buckle you into the cockpit of his airplane. He will bury you deep in blue engine smoke or in the soft live silence of soaring flight; he will take the world in his hands and twist it all directions before your eyes. He will touch a machine of wood and fabric and bring it alive for you; instead of seeing speed from a cabin window, you will taste it in your mouth and feel it roaring by your goggles and watch it fraying your scarf in the wind. Instead of knowing height on the dial of an altimeter, you will see it as a tall, wide air-filled space that begins at the sky and drops right straight down to the grass. You will land in hidden meadows where no man or machine has ever been, and you'll soar upslope on a mountain ridge from which the snow sifts downwind in long misty veils.

You'll relax in soft armchairs after supper, in a room whose walls are covered with airplane pictures, and feel the thunder and shock of ideas and perfection surge like a hurricane sea over the faces of skill around you. The sea calms near sunup, and the Flyer drops you off at home in the morning ready only to fall into bed and dream of airfoils and precision flying, thermal-sniffers and racing in the ground effect. Great suns roll through your sleep, and colorful checkerboard land drifts below.

When you wake you might be ready to make a decision one way or another, for Aviation or Flying.

Rare is the man who has been exposed to the intense heat of a pilot's enthusiasm, without being in some way affected by it. The only reason that this can be is the unreasonable itself, that strange distant mystique of machines that carry men through the air.

Aviation or Flying, take your choice. There is nothing in all the world quite like either one of them.

The author of *Fair Stood the Wind for France*, one of the finest novels to come out of World War II, Herbert Ernest Bates was himself a flight lieutenant in the Royal Air Force. In this moving piece he conveys, with the particularity of vision that is his trademark, the nobility of an all too common death.

H. E. BATES

It's Just the Way It Is

November rain falls harshly on the clean tarmac, and the wind, turning suddenly, lifts sprays of yellow elm leaves over the black hangars.

The man and the woman, escorted by a sergeant, look very small as they walk by the huge cavernous openings where the bombers are.

The man, who is perhaps fifty and wears a black overcoat and bowler hat, holds an umbrella slantwise over the woman, who is about the same age, but very gray and slow on her feet, so that she is always a pace or two behind the umbrella and must bend her face against the rain.

On the open track beyond the hangars they are caught up by the wind, and are partially blown along, huddled together. Now and then the man looks up at the Stirlings, which protrude over the track, but he looks quickly away again and the woman does not look at all.

"Here we are, sir," the sergeant says at last. The man says "Thank you," but the woman does not speak.

They have come to a long one-storied building, painted gray, with Squadron Headquarters in white letters on the door. The sergeant opens the door for them and they go in, the man flapping and shaking the umbrella as he closes it down.

The office of the Wing Commander is at the end of a passage; the sergeant taps on the door, opens it and salutes. As the man and the woman follow him, the man first, taking off his hat, the woman hangs a little behind, her face passive.

"Mr. and Mrs. Shepherd, sir," the sergeant says.

"Oh yes, good afternoon." The sergeant, saluting, closes the door and goes.

"Good afternoon, sir," the man says.

The woman does not speak.

"Won't you please sit down, madam?" the Wing Commander says. "And you too, sir. Please sit down."

He pushes forward two chairs, and slowly the man and the woman sit down, the man leaning his weight on the umbrella.

The office is small and there are no more chairs. The Wing Commander remains standing, his back resting against a table, beyond which, on the wall, the flight formations are ticketed up.

He is quite young, but his eyes, which are glassy and gray, seem old and focused distantly so that he seems to see far beyond the man and the woman and even far beyond the gray-green Stirlings lined up on the dark tarmac in the rain. He folds his arms across his chest and is glad at last when the man looks up at him and speaks.

"We had your letter, sir. But we felt we should like to come and see you, too."

"I am glad you came."

"I know you are busy, but we felt we must come. We felt you wouldn't mind."

"Not at all. People often come."

"There are just some things we should like to ask you."

"I understand."

The man moves his lips, ready to speak again, but the words do not come. For a moment his lips move like those of someone who stutters, soundlessly, quite helplessly. His hands grip hard on the handle of the umbrella, but still the words do not come and at last it is the Wing Commander who speaks.

"You want to know if everything possible was done to eliminate an accident?"

The man looks surprised that someone should know this, and can only nod his head.

"Everything possible was done."

"Thank you, sir."

"But there are things you can never foresee. The weather forecast may say, for example, no cloud over Germany, for perhaps sixteen hours, but you go over and find a thick layer of cloud all the way, and you never see your target—and perhaps there is severe icing as you come home."

"Was it like this when—"

"Something like it. You never know. You can't be certain."

Suddenly, before anyone can speak again, the engines of a Stirling close by are revved up to a roar that seems to shake the walls of the room; and the woman looks up, startled, as if terrified that the plane will race forward and crash against the windows. The roar of airscrews rises furiously and then falls again, and the sudden rise and fall of sound seems to frighten her into speech.

"Why aren't you certain? Why can't you be certain? He should never have gone out! You must know it! You must know that he should never have gone!"

"Please," the man says.

"Day after day you are sending out young boys like this. Young boys who haven't begun to live. Young boys who don't know what life is. Day after day you send them out and they don't come back and you don't care. You don't care!"

She is crying bitterly now, and the man puts his arm on her shoulder. She is wearing a fur, and he draws it a fraction closer about her neck.

"You don't care, do you? You don't care! It doesn't matter to you. You don't care!"

"Mother," the man says.

Arms folded, the Wing Commander looks at the floor, silently waiting for her to stop. She goes on for a minute or more longer, shouting and crying her words, violent and helpless, until at last she is exhausted and stops. Her fur slips off her shoulder and falls to the ground, and the man picks it up and holds it in his hands, helpless, too.

The Wing Commander walks over to the window and looks out. The airscrews of the Stirling are turning smoothly, shining like steel pinwheels in the rain, and now, with the woman no longer shouting, the room seems very silent, and finally the Wing Commander walks back across the room and stands in front of the man and woman again.

"You came to ask me something," he says.

"Take no notice, sir. Please. She is upset."

"You want to know what happened? Isn't that it?"

"Yes, sir. It would help a little, sir."

The Wing Commander says very quietly: "Perhaps I can tell you a little. He was always coming to me and asking to go out on operations. Most of them do that. But he used to come and beg to be allowed to go more than most. So more often than not it was a question of stopping him from going rather than making him go. It was a question of holding him back. You see?"

"Yes, sir."

"And whenever I gave him a trip he was very happy. And the crew were happy. They liked going with him. They liked being together, with him, because they liked him so much and they trusted him. There were seven of them and they were all together."

The woman is listening, slightly lifting her head.

"It isn't easy to tell you what happened on that trip. But we know that conditions got suddenly very bad and that there was bad cloud for a long way. And we know that they had navigational difficulties and that they got a long way off their course.

"Even that might not have mattered, but as they were coming back the outer port engine went. Then the radio transmitter and the receiver. Everything went wrong. The wireless operator somehow got the transmitter and the receiver going again, but then they ran short of petrol. You see, everything was against him."

"Yes, sir."

"They came back the last hundred miles at about a thousand feet. But they trusted him completely, and he must have known they trusted him. A crew gets like that—flying together gives them this tremendous faith in each other."

"Yes, sir."

"They trusted him to get them home, and he got them home. Everything was against him. He feathered the outer starboard engine and then, in spite of everything, got them down on two engines. It was a very good show. A very wonderful show."

The man is silent, but the woman lifts her head. She looks at the Wing Commander for a moment or two, immobile, very steady, and then says, quite distinctly, "Please tell us the rest."

"There is not much," he says. "It was a very wonderful flight, but they were out of luck. They were up against all the bad luck in the world. When they came to land they couldn't see the flare path very well, but he got them down. And then, as if they hadn't had enough, they came down slightly off the runway and hit an obstruction. Even then they didn't crash badly. But it must have thrown him and he must have hit his head somewhere with great force, and that was the end."

"Yes, sir. And the others?" the man says.

"They were all right. Even the second pilot. I wish you could have talked to them. It would have helped if you could have talked to them. They know that he brought them home. They know that they owe everything to him."

"Yes, sir."

The Wing Commander does not speak, and the man very slowly puts the fur over the woman's shoulders. It is like a signal for her to get up, and as she gets to her feet the man stands up too, straightening himself, no longer leaning on the umbrella.

"I haven't been able to tell you much," the Wing Commander says. "It's just the way it is."

"It's everything," the man says.

For a moment the woman still does not speak, but now she stands quite erect. Her eyes are quite clear, and her lips, when she does speak at last, are quite calm and firm.

"I know now that we all owe something to him," she says. "Good-bye."

"Good-bye, madam."

"Good-bye, sir," the man says.

"You are all right for transport?"

"Yes, sir. We have a taxi."

"Good. The sergeant will take you back."

"Good-bye, sir. Thank you."

"Good-bye," the woman says.

"Good-bye."

They go out of the office. The sergeant meets them at the outer door, and the man puts up the umbrella against the rain. They walk away along the wet perimeter, dwarfed once again by the gray-green noses of the Stirlings. They walk steadfastly, almost proudly, and the man holds the umbrella a little higher than before, and the woman, keeping up with him now, lifts her head.

And the Wing Commander, watching them from the window, momentarily holds his face in his hands.

Here we have one Canadian ace reporting on the techniques and heroics of another, the man he considers "the greatest air fighter the world has ever known." Billy Barker was a wild man but hardly a fool; he rightly assessed the relative weaknesses and strengths of the Sopwith Camel and Snipe and attacked from strength. Bishop was considered no less a misfit than Barker. During their training periods, both displayed a cavalier attitude toward their planes — and, indeed, their lives — that made them the most unlikely candidates for sustained success in combat. Yet both possessed that combination of daring and doggedness that secured their place in aviation history — Bishop with seventy-two victories, trailing only Mick Mannock by one and René Fonck by three among Allied airmen, and Barker with fifty-three.

WILLIAM A. BISHOP

Billy Barker's Brand of Attack

He was a Canadian, and his name was Billy Barker. Like so many North Americans who have left their imprint on the pages of combat, he came from a midwest prairie farm and had no experience of war, or aviation, and little of the world itself, when he joined the Canadian forces in 1914 as a private soldier. In this respect at least he bears a marked resemblance to America's young Major Bong. It was 1916 before Barker transferred from the infantry into the old Royal Flying Crops, which he joined, like myself, as an observer, in which role he had more experience than I, for by the time he went back to England from France to learn to fly he had already been wounded and had won the Military Cross.

Billy Barker went back to the war as a pilot early in 1917, this time as chauffeur of a two-seater artillery observation machine. After a tour of operational duty at the front he was sent back to England to be an instructor.

Barker felt completely frustrated by living in a Training Squadron, teaching other youngsters how to fly. He applied to be returned to operational duties, but the application was denied. Whereupon he took a most unmilitary and undisciplined step and decided to make himself such a nuisance to the staff that they would be glad to see the last of him. Day after day he was pulled up on the carpet for low flying over towns, for stunting over headquarters, and for playing the fool generally. Called to account by his superiors he promptly informed them that he intended to misbehave and commit compound fractures of the rules until such time as the brass hats decided to post him back to the front.

So Barker was sent back to France by his irritated superiors, this time to fly a Sopwith Camel, that delicate piece of sudden death with which British pilots wrought so much havoc amongst the Albatroses in the later months of the war.

Barker was the originator of one great trick. That was to lure his enemies into battle as close to the ground as possible, for he had made the discovery that the Camel was infinitely more maneuverable close to the ground than any airplane the Germans possessed.

The first great test of the Barker system came one day at dusk when he was leading his flight home, but was still deep in German-held territory. Suddenly ten planes appeared in the half-light heading east: German Albatroses on their way home. Barker and his friends went in to attack.

Early in the melee the Canadian pilot, while going to the assistance of one of his mates, found himself with a Hun on his own tail. This was his chance to try out the system. Zigzagging to keep out of range of the German's guns, Barker circled lower and lower, making no attempt to get into shooting position, coaxing the German pilot right up the garden path. He led the enemy almost down to the treetops, then suddenly — and suddenly is altogether too slow a word — went into a tight loop with his small, quick Camel. The slower Albatros could not cope with such activities. In a split second Barker was on the German's tail and with one quick burst destroyed him.

Almost before the first had crashed, a second pounced on Billy. Using the identical tactics he crashed the second German in flames, zoomed away, and made for home in the falling light. Such a master of low fighting did Barker become in the next few weeks and such a madcap devil was he in his sheer youthful exuberance that Headquarters hit on a brainy idea and formed a squadron of youngsters who were then trained to the Barker system and shipped to Italy to fight the Austrians and Germans in the mountain passes.

In Italy Barker's bunch ran wild. They blew Austrian captive balloons, hitherto unmolested, out of the sky far behind the lines. They strafed enemy aerodromes at grasstop height, pouring inflammable bullets into the open doors of hangars to set them aflame. They played particular and general hell with the easygoing Austrians.

The enemy, including German reinforcements sent to the Italian front to deal with this mad Canadian and his lunatic youngsters, tried bombing Barker's aerodrome with more than thirty German Gothas and on the first try lost almost half their force when Barker and his boys soared up into battle. Barker had now developed a new brand of attack much used in World War II. It is called the head-on system and it takes plenty of guts to be properly executed and get your man, because you must fly at him on his level, straight down his path toward him, and keep going until somebody weakens. I cannot remember ever having met a German who did not duck first. They certainly used to duck when they saw Barker coming.

That was Billy's way of dealing with the Gotha bombers, and on his first encounter he took two out of three by this device. That, in my opinion, must have been one of the great shows of the war. The three Germans were flying in line astern and Barker flew head-on to meet them. He nabbed the first when they were almost bow to bow, ducked under the second, came up head-on into the third, knocked it out of formation, turned on it, and plunged the big Gotha to earth a flaming wreck.

But it was in low flying where he always excelled. Most pilots preferred the more formal method of getting up high, between the enemy and the sun, and pouncing on him from that vantage point. Barker wanted them down near the ground and to find them near the ground meant he practically had to visit their aerodrome and lie in wait. His superiors considered him completely mad, but the system worked and soon Billy was knocking enemy aircraft down like flies. Decorations came his way one after the other. The D.S.O. was soon added to the M.C., won as an observer. Then came a bar to his D.S.O. and a bar to his Military Cross. Early in May 1918 the Italians decorated him and Barker subsequently confided in me that, while it was nice to be honored by one's Allies, it was *not* nice to be kissed on both cheeks by a man with a beard.

Pehaps one of the funniest stories about this terrific fighting man is what his friends call the Spy Story.

The practice had come into vogue of flying Allied spies across the enemy lines and dropping them by parachute to do their jobs. At that time, of course, very few people knew much about chutes and extremely few thought highly of dangling under the silk to descend to the ground in the middle of the night. Consequently, it was often the case that a spy would set off in the rear cockpit of an aircraft to be dropped at an appointed place by the pilot, but prior to arrival would achieve understandable frigidity of the feet and decide not to make the jump. To meet this problem Barker, who had had a couple of experiences with mind-changing spies, devised his own system, which was simply to equip the passenger seat with a trapdoor arrangement operated by a lever from the pilot's cockpit, the passenger being totally unaware of the arrangement. Then Barker would fly off with his spy for delivery behind the Austrian lines, keeping the nervous gentlemen comforted with reassuring noises as they flew along. When the appointed destination was reached Billy simply sprang the trap. The spy fell out through the floor and died a thousand deaths while he waited for his parachute to open. Inasmuch as each passenger traveled on a one-way ticket, it was not difficult to keep the system secret from prospective riders. Thus for some time Billy operated a spy delivery service which won him high repute with the Intelligence staff.

The greatest Barker story of all relates to three famed Austrian pilots with whom Billy and his teammates sought combat but who had consistently ducked whenever Barker and his Camels appeared on the scene. They were looking for sitting birds, nice quiet artillery observing two-seaters, not maniacs who seemed to be able to do anything they liked with their midget fighting machines.

So Barker had thousands of leaflets printed, and dropped them behind the Austrian lines. The leaflets carried the following challenge:

Major W. G. Barker, D.S.O., M.C., and the Officers under his command present their compliments to

Captain Brumowsky,
Rither von Fiala,
Captain Havratil,

and the pilots under their command and request the pleasure and honor of meeting in the air. In order to save Captain Brumowsky, Rither von Fiala and Captain Havratil and the Gentlemen of their party the inconvenience of searching for them, Major Barker and his Officers will bomb Godega Aerodrome at 10:00 A.M. daily, weather permitting, for the ensuing fortnight.

Barker and his young men carried out their schedule to the letter and the moment. They bombed Godega daily through the fortnight. Once or twice the enemy appeared. But they were face-saving appearances and Barker, as always, came out on top by a lopsided score.

Billy left Italy in the summer of 1918 and returned to Britain to pursue a new course in air-fighting tactics—a young man off to school to learn from the book he had practically written. As always he rebelled against the discipline and red tape of the war behind the lines. This time he evolved a new excuse to get back into combat and asked to be allowed a few weeks at the front in France before taking the course, to acquaint himself with new German tactics. There, on October 27, 1918, only two weeks before the Armistice, and on the flight which was supposed to take him back to Britain, he put on a show which stands as the greatest in the annals of World War I, and one which it would be hard to tie in the later to-do.

The official account states that Barker, having packed his kit and seen it shipped to England, climbed into the Sopwith Snipe he was to fly back, but couldn't resist taking one more look at the lines. He decided to do a little wandering and to have one last look at the war he was on the point of leaving. Instead of turning toward the Channel he swung off over the lines. In a few moments he saw a German machine at 22,000 feet and attacked. The German observer fired so accurately that the Canadian's own machine was badly hit. As Barker's journey was to have been routine and peaceful, the telescopic sight had been stripped off his guns before taking off, so Billy was equipped with ordinary peep-sights, nothing else. Twice he attacked the German before he was able to kill the observer. Then he closed in to short range and shot the pilot and airplane down. The enemy machine broke apart in the air and fell in a rain of small pieces. At this juncture a Fokker pounced on Billy, catching him by surprise, putting an explosive bullet into his right thigh and shattering it. That would have been enough to cause any ordinary fighting man to get out of there, but not Billy Barker. Billy was going to get even. He stayed with his German, finally brought him into the peep-sight for a split second, and sent him reeling down in flames.

Still Billy had not enough, although by then he was fighting to retain consciousness. He must have passed out because, as he told me afterwards, he suddenly found himself in the midst of a crowd of enemy machines, the number of which was estimated by people on the ground as being at least sixty, without having the slightest idea how he got there.

Germans jumped him from every corner of the sky. His machine was hit repeatedly and he was severely wounded again, this time in the left thigh. He fainted from loss of blood and fell thousands of feet out of control with the whole German circus after him. The rush of air in his wild dive brought him back to life and suddenly he turned on his enemies like a mad dog. By then any hope of survival must have gone from his mind. He was simply going to wreak all the havoc possible before the enemy fliers polished him off. He charged head-on at an enemy machine, thinking to collide with it and take at least one more German with him where he was sure he was going.

But, as always, his bullets were right on the target. Before the collision could occur the German burst into flames and fell out of battle—and Barker had picked up an explosive bullet in his left elbow. To tally the score at this juncture, Barker was now sitting in his cockpit, with one thigh shattered, the other severely wounded, and his left arm limp and useless. He fainted again, and again fell out of control. Again he recovered and again swooped up into the melee. This time he fought the Germans all the way down the long hill almost to the ground and in the course of the battle shot down two more. By then he was close to the ground and still under attack from many German machines. A burst of explosive bullets perforated the gasoline tank under his seat, but by miraculous good luck—and there were no self-sealing gas tanks then—the machine did not take fire. Barely conscious, Barker switched to his auxiliary tank and kept the little Clerget rotary engine spinning. He fainted again and almost spun in, recovering consciousness and pulling out just in time. With the machine almost out of control he put its nose down and headed west, not knowing where he was, and piled his machine into a maze of barbed wire immediately behind the British lines. Five German aircraft had gone down in that tremendous melee between one man and God knows how many antagonists, a man who was supposed to be quietly flying across the Channel to Britain to take a course in air fighting! They gave Billy the V.C. for that one.

His wounds mended and soon after the end of the war he returned to Canada, where he and I were happy partners in one of the first and most amusing commercial aviation enterprises ever undertaken by foolish young men. The irony of life caught up with Billy Barker soon afterwards. He died by stalling and spinning into the ground just after taking off from Ottawa in one of the first Fairchild machines. So passed a man who, in my book, stands as the greatest air fighter the world has ever known.

On his way to becoming America's foremost pilot of experimental ships like the Skyrocket in the mid-1950s, Bill Bridgeman, a wartime pilot of four-engine behemoths, had to go to school again. In this excerpt from his 1955 book, *The Lonely Sky*, Bridgeman reports to work at Douglas Aircraft and prepares for the AD, a single-engine attack plane. He will test it; it will test him.

WILLIAM BRIDGEMAN and JACQUELINE HAZARD

Education of a Test Pilot

At the field I gave my name to the cop on the gate and was allowed to pass. Laverne Brown, the man I was to see, was sitting at a desk in a narrow little room labeled Country Club Office. He was older than I thought he would be, around forty maybe, tall, leanly built, and dark. His skin was brown and weathered. I introduced myself.

"Hey, I'm glad to see you." The way he lit up I thought he really meant it. He waved his arm in the direction of the field where the attack planes were lined up with their wings folded. "I could sure use some help around here, they've been piling up lately."

Despite all the newly-put-together, never-been-tried AD's crowding the field, production wasn't exactly booming at Douglas. The entire testing staff at El Segundo consisted of Laverne Brown and now me. And there weren't more than ten engineering pilots scattered throughout the other five Douglas plants. The manufacturing of airplanes wasn't one of the more "going" industries after the war.

Brownie showed me around, introduced me to Jerry Kodear, the dispatcher, and a couple of inspectors.

"Take a locker. We'll get you something to fit you tomorrow." He reached into his desk and pulled out a handbook on the AD. "Look this over . . . tomorrow we'll try it out."

It had been a long time since I had been in a single-engine ship, not since my days of ferrying fighters for the Navy, almost four years ago.

The next morning came up, a hot August day. By eight-thirty I was in the flight room having a cup of coffee and going over the AD manual once more when Brownie showed up.

"Think you can find all the knobs?" he asked.

"I think so. A few minutes in the cockpit will help."

"We'd better go over the starting procedure together. Those Wright engines can be rough if you've never been behind one before." Brownie had the assured air that a man who has been flying for over twenty years wears. He knew more about how to handle a plane than any pilot I had ever met. The AD before me was the first I had seen at close range but Brownie didn't take advantage of that fact. He left me pretty much alone.

The panel came as advertised in the handbook. Brownie leaned into the cockpit, ready to give me a hand getting her started. With a deftness that comes with 1,000 hours in a ship, he brought the big engine to life. I watched the fast sequence of his motions, trying to memorize them.

Over the clattering roar of the engine he shouted, "If anything goes wrong phone us. Take her up for a half-hour or so and get acquainted." He was through; the ship was mine. There was no fatherly advice. He jumped down and walked back toward the hangar.

Now I tried to remember everything I could about single-engine airplanes. *Remember, with the power you've got here it's going to take a lot of rudder to hold this thing on the runway.* What else was there to remember?

Brownie wasn't kidding, the engine sounded as if it was ready to come out of the mounts. I pulled into the run-up position, unfolded the wings, checked the mags. A quick look at the engine instruments and I was cleared for takeoff. On two-five-right, the Los Angeles Airport companion runway, Southwest was starting her takeoff roll. Friends from home. I made the senseless gesture of waving and then self-consciously pulled my hand down. Southwest ignored me as she lumbered by.

"She's got a high rate of climb," Brownie had warned me. I lifted her off and pointed up. *A high rate of climb?* The AD went upstairs like she was shot in the fanny.

At 18,000 I nose over, reduce power, and there it is again—the world, new, and again always the same burst of exhilaration on the first look down, like Monday-morning flight after a long weekend. Here is a kind of freedom that I find nowhere else. You are on top of it, nothing can hide from you, an explorer in an empty sky. Up here you're big. You can move.

I turn the energy churning in front of me south toward Laguna. There are no passengers in back of me and there is no schedule, no place to be, just me and the AD and the coast of California edged by the mountains. The attack plane is more power than I have ever handled; it's a big engine, a small airplane. At Laguna I run out to the ocean, turn north, and jump the waves. The AD is a colt in a blue meadow.

Into a steep climb, the engine kicking up a storm, I pick up altitude and level off. It has been a couple of years since I have been on my back. I try a few rolls and some Immelmanns. I dish out of the rolls and the Immelmanns are sloppy. Some old fighter pilot down there is probably staring at the sky shaking his head.

The dispatcher calls, asking my position. I had neglected to call him and he is checking to see if his new boy is lost. I report my position. The half-hour is gone and on the way back to the field I check Channel 3 on the radio and the remaining two items that had been given me to test on this first flight.

My confidence is somewhat shaken by the sloppy way I have handled the ship in the rolls, and the landing before me I approach with caution. There is another audience, for sure, at the hangar. I keep a little more power on than usual and my approach is fast. I allow lots of room; this time I don't try and see how close I can come to the end of the runway at touchdown. A nice, easy landing. There is no one watching.

Brownie greeted me in the office. "How did it go?"

"Noisy Goddammed things, aren't they?"

Brownie snorted, "You'll think so when you get a rough one. You know the area fairly well, don't you, Bill? Now I suggest you memorize the radio frequencies of all military fields so you can warn them if you've got to set down in a hurry." He went on in a serious voice. "We make it a rule that everybody wears a Mae West and carries shark repellent." He caught the side look I gave him. "It's up to you, boy, we all wear it. And another thing. Try to call in every twenty minutes or so; Jerry likes to know where the planes are." Then Brownie smiled. "With Jerry we like to kid a little. Pick out a spot on the map that nobody has ever heard of. It drives Jerry crazy trying to find it. Everybody does it to him." Then he continued, serious again.

"If you get in trouble, let me know. If I'm on the ground I'll try to help you out; if I'm in the air I'll join up and give you a hand. We'll let you take it easy these first couple of weeks until you get to know the ship."

My first two weeks as test pilot I was allowed to take up second and third flights, after Brownie had worked the planes over first. While I was furtively trying to execute a respectable roll in a hunk of vacant sky over orange-grove country 40 miles from the field, after I had completed my test items, Brownie was taking up "first flight" on six to seven planes a day. He would set one down, make out a squawk sheet, climb into the next one, go over the items to be tested, bring it back, write it up, stop for coffee, and off he would go again. It took him little over a half an hour to run through the check items on each ship. He handled the AD like a kiddy car.

Brownie's Immelmanns and rolls were precise and unfaltering; I didn't mind confessing to him that I was having trouble with mine. And although Brownie wasn't one give to "hangar flying" he explained in great and careful detail how to keep from dishing out. After twenty years in the air Laverne Brown, who once portrayed the dashing

IN THE YEARS AFTER WORLD WAR II COURAGEOUS TEST PILOTS LIKE

Bill BRIDGEMAN

FLEW AMERICA INTO THE AGE OF SUPERSONIC SPEED.

IN 1951, BRIDGEMAN PILOTED THE EXPERIMENTAL DOUGLAS D-558-2 "SKYROCKET" TO NEARLY TWICE THE SPEED OF SOUND.

"Tailspin Tommy" in a move serial some years back, was a little bored with flying. He had long since outgrown the wild-blue-yonder stage, but he saw with amusement I still liked kicking it around. Flying continued to be a form of sport with me.

And now I liked what I was doing after the confining routine of the airlines — getting up there and breaking loose. In testing as in no other branch of flying you are on your own; no one is leaning over your shoulder. And even though, again, it seemed like a temporary job, with little possibilities for the future, I was happy.

At the end of two weeks Brownie gave me a never-been-off-the-ground AD to test. "These things have got a sensitive carburetor, Bill, on first 'go' you have to find out quick if it's correctly set. Take it up to 30,000. It'll cut out and backfire if the setting is off. You'll probably get a warning before that — the engine gets rough." He handed me a stack of cards. They listed power-plant tests to be made: a stall-warning chart to be filled, an autopilot card with ten items to be noted and a bunch of miscellaneous checks. The next-to-last item on the last card included climb, dive, and banking maneuvers. *Brownie knocks this stuff off in forty minutes?*

"Circle and stay close to the field so you can put 'er down quick if she cuts," Brownie finished.

A new plane smells like nothing else: a man-creation with its own odor-mixture of newly jointed, polished parts. The AD was spotless and ready to be tried, the original model had been tested by the engineering pilots for design faults, for stability and control, to see if it would fly. It had been modified and passed and the El Segundo plant was stamping them out at a rate of two a day. Now all that remained was to see if each airplane, as it rolled onto the field, had its parts put together properly. A thousandth of an inch off and the screw would cause the ship to complain and protest.

Walking up to the ship — a plane that had never been in the air — her wings folded as if she were shrinking from her purpose, a bizarre thought occurred to me. Maybe this one won't fly at all, the parts won't mesh. No one has ever proved this one.

This time I knew Brownie had his eye on me. I read over the cards carefully. Each card had items to be checked at varying altitudes, knobs to be turned at different speeds, and items to be tested in between trim alterations. There would be no sight-seeing on this flight; this was going to be work.

Before I wind it up I enter the gauge readings and set the knobs for items to be checked at takeoff. The big Wright engine rips up the morning and I taxi out on the runway. She moves . . . now let's see if she flies. The AD picks up over the field, points high. She flies! Things are working for the first time, the parts are circling, pumping, meshing, as they have been engineered to do. A minor miracle.

Climbing to 24,000 to 26,000, to 28,000 feet. I suck my air from a bottle of oxygen. The air is thin outside and the carburetor now must feed the engine its fuel — neither too lean nor too rich. It is set correctly. She doesn't get rough and I turn to the cards clipped to the board that is strapped to my knee. As I come to each new card with its series of tests, I drop or climb to the altitude that the item requires. The entire flight is a matter of reaching for knobs, pulling switches, checking the radio, climbing up, and diving all over the sky. Two hours later I return the ship to Brownie: almost four times as long as it takes him for the same operation.

My arms and legs ached from all the activity in the new attack ship and wearily I unfolded into a chair with coffee and the squawk sheet that remained to be filled out — the pilot's literary attempt at evaluating the ship, an aviation critique.

Brownie came in from a flight, a parachute dangling from his back. "How did it go?" He asked for the cards and my squawks. "You think she needs a little nose-down trim at cruise power, huh?"

"Another half-degree will do it. God, I'm knocked out. How many operations can a plane have . . . up and down, back and forth, pull this, pull that. . . ."

He read the cards. "You'll get used to it. Next time you'll coordinate your items as to altitude — it saves time."

He looked at his watch and went out on the field again. I finished my coffee and followed after him. From the hangar door I could see him taxiing down the field in Number 48, the ship I had just brought in with eight squawks against it. Forty minutes later he landed. I waited for some comment as he headed for the flight office. He walked slowly. It was past quitting time; when he passed me he didn't stop. Over his shoulder he said, "See you tomorrow, Bill." He smiled.

First flights became easier; I learned to group items under altitude headings. At 5,000 I would get everything out of the way that had to be tested at that altitude, the same at 10,000, and all the way up. Brownie never checked another plane of mine until the day I brought one back with a clean bill of health. Nothing was wrong with it. "I'll buy it, Brownie. As far as I'm concerned she's sold." Every airplane that is delivered bears the name of the pilot who "bought it," who okayed it as a satisfactory working unit. If it comes up with a defect, the Navy refers back to the man who tested it. The test pilot is responsible for the performance of the plane.

Later that day Brownie took my no-squawk plane up for a flight. He never said a word. The report stood. The plane was "sold." I was a test pilot and it was a nice pleasant way to earn a dollar.

After I got the procedure organized, I could handle four to five planes a day. On first flight I no longer hovered over the field. I flew down toward Laguna, inland over the square acres of dark green orange trees; an area not too heavily housed, close to a couple of military fields and not many miles form the field at El Segundo. It became my ground.

"KAK6 calling. I'm over Anaheim, Azusa, and Cucamonga."

"Ah, knock it off, Bridgeman," Jerry's voice comes into the cockpit.

The ship is "sold" on this flight. She had been in good shape to start with, only six items against her and in three flights they were worked out. I am bringing her back for delivery. It is Monday morning and it is clear and sparkling and exhilarating, like the first time up, the first time you solo. There is the boundaryless sky-country to run in.

Grinding up the coast over Balboa admiring the blue and white sunlit day, I have no desire to be in the water below me. I'm water-soaked from the weekend. It is good to be flying again. My glance sweeps past the sun; it picks up what appears to be a speck in the sky. The sun is too bright to investigate. A part of a moment later a flash of aluminum goes by and the bright blue Skyraider wallows in the passing prop wash. The sudden intrusion shakes me. I'm a little annoyed and I search the sky for my new playmate. Off to the right and below, the sun that hid him now gives him away as momentarily the reflection bounces off his wings. It's a Mustang pulling out of a dive, clawing for altitude. The energy he picked up in the dive will take him well above me. *Race you to the corner* . . . and the fight begins.

Instinctively, in one motion, I catch the throttle and prop controls and sweep them all the way forward. She groans a little, shakes herself, and leaps forward. I pull the nose up and hang it there. The big prop chews up air and throws it behind; Balboa falls back far below. Above me the Mustang turns and starts his second pass. I turn slightly away. He turns in and commits himself. I didn't expect him to bite that quick. This is the pass he'll use. I turn and point the blue nose straight for him. In a wild gyration he swings to avoid me. The maneuver equalizes our positions. We stand off at approximately the same altitude and speed, two great engines churning up the quiet morning over beach and ocean. The giant circle begins. The turn into it puts three G's on me. The stalking. Pulling the circle tighter and tighter. The tight turns pick up G fast; I pull to the edge of stall and hold it there. She shakes and tends to roll off on a wing, I ease off the G a bit. The force of the turns pulls the skin of my face down tight and grabs my mouth open. As the blood is drained out of my brain, I can feel the blackout coming on; the world turns a foggy gray. I shake my head and scream from my bowels to keep the blood from pooling there and in my legs. I am used to pulling G's in flight tests, I can stand up to six without a "suit." The gray fades away now and I look over at my friend to see how he is doing. He no longer is directly opposite me in the circle, my turn is a hair less the diameter of his. Degree by degree the Skyraider is closing in on him, gradually I am moving up behind him. Were this combat, the process of "slow death" would have begun. In a moment it would all be over. Now his futile evasive action begins. He flips into a violent split S and shrieks off in a long, steep dive for the water and beach below. I flip with him, right behind him for a short time, and then pull up wing to wing. The Air Force fighter pilot looks over, nods his head, and smiles. Very slowly he rolls his airplane. Clumsily, I try to follow. Here I am no match with a fighter pilot. I know, too, if the ships had been switched, I would have been no match in the big circle above.

"KAK6, this is 482, I'm over Yapica and I'm coming in."

"You're late for lunch, boy."

There were three more planes waiting for me after lunch. The last one was a first flight.

The engine sounds smooth enough until we get close to 20,000, then suddenly the regulator controlling the carburetor isn't doing the job. The ship begins to shake.

Well, here it is. This is the day, the day they are paying you for. I become alert as a man who has had a bullet go by his ear. I pull the throttle back, but not fast enough. W-hop! Bang! The ship trembles under the blast. Brownie's carburetor threat has arrived. The AD drops through 3,000 feet. At the lower altitude, where the air is thicker, I optimistically apply power, the engine protests in violent grinding and shuddering. The ship is not going to make it back, the engine won't hold up and at once panic that is begining to crowd in me stands still for a second—this is what I am hired to do. Get the plane back. My ego and a separate calm consciousness that sees the practical and thrifty side of the issue prevent the basic reaction to jump from taking hold. Panic will have to wait.

Make up your mind quickly. Which field is closest? El Toro, Long Beach, Alameda? It had better be the right decision; with the loss in power the ship is settling fast. Five minutes ago things were in their proper place, a slight change in adjustment and now the once-obedient machine is 12,000 pounds of trouble. Long Beach, El Toro, and Alameda, warm-sounding places, might as well be a million miles away—there isn't enough power left to get to a field. Painfully I repeat a four-letter word into the metal-cold box I sit in. The damn thing won't make it to a field. A spot to set her down. There. A wide place in the road—not big enough. It is swept under and behind me. Over there, that brown field. That has got to be it. The ground is moving up closer beneath me. The field is covered with tall weeds that hide level ground or maybe deep holes or a pile of rocks . . . idle speculation, because there is no choice. One last attempt at the throttle. The AD coughs a response and the power increases a hair. Enough to get to a field? To leave it here is awkward; it would have to be towed home. I will nurse it back on the power it has left; it is better to bring the ship home myself than to have the company tow it back down the highway. I can do it.

The engine protests; it does not want any more and it shudders in its mounts as if it were trying to tear loose and drop to the ground.

"KAK6 . . . 522 returning to field. This is an emergency. Over."

The reply is immediate: "Five-two-two . . . do you want emergency equipment to stand by?" Jerry's voice is calm and precise.

"Yeah, I think you better." A game with the dispatcher—who can sound the most detached? Under me the even squares of houses are getting thicker with their back yards of fenced-in kids. Just a little longer, Goddamm it, a little bit longer. The houses close beneath are running away from the clattering, sick, blue plane that rushes over them. Come on, baby! I suck in my breath, begging her home. My position is low. The engine could give up any minute and there is no altitude to glide in from and no brown field below to head for.

"Five-two-two . . . what is your position now?" The voice intrudes.

"Never mind," I snap. They don't ask again. I haven't time to answer questions. With each poke of the throttle she rattles and moans. Oh, Christ, come on, come on!

"Five-two-two, we have you in sight. You are cleared to come in on any runway. . . ."

And there it is, the field, I get one chance at it, I won't be able to come around again.

The gear touches the ground. Tranquil, serene, and utterly beautiful field. I lean back in the seat and sop up the delicious comfort of just sitting. The harassed airplane under me bleeds oil and pants smoke. Then two firemen are on the wings, pulling me out of the ship. I resent the intrusion.

"What's the matter? She won't go up," I shout at the men. They haul me out anyway and the ambulance attendants move quickly over to us. I wave them away. "There's nothing."

In the hangar I dumped my chute in my locker and headed for the shower, the collar of my overalls was wet with sweat.

"Holy God," somebody complained from the radio room, "I never would have tried to get the thing in. You would be nuts to try."

"It sure is a mess all right. Anything to be a hero, huh?"

I was dressed and writing out a report when a mechanic walked in.

"What does it look like?"

The mechanic looked at me in disbelief. "It looks like hell. You broke three push-rods, you seem to have swallowed a valve, and number-three jug is completely disintegrated."

"Well, it's home anyway."

"Yeah, it sure is."

This time I was wrong. I had brought it in but I had not won the contest. I had made an error in judgment. I was wrong for a second when I made up my mind to nurse it back. A block, any block of those small, square houses along my path, just escaped having an engine bowled through a living room or a fire that could have eaten away its whole lapful of homes.

Next time there will be no decision with an engine that rough. It will merely be a set of circumstances that has only one answer: set it down quick.

In ten nasty minutes, bringing the strangling AD home, the first three weeks of freedom were paid for.

In 1927 *Wings* won the first Oscar given for best motion picture. Hundreds of aviation films have come since, but none has equaled it in conveying the feeling of *being* in aerial combat. The dogfights, the mission against a pair of dirigibles, the Spads spinning to their fiery deaths—these are not merely gripping filmmaking; the action is so authentic, so meticulously rendered, that the film has become for modern viewers virtually a documentary of World War I in the air. The film is occasionally revived in television retrospectives of the silent era, but almost never is it shown in a theater, on a proper-size screen. Any armchair aviator who chances to see *Wings* in this setting will never forget it.

KEVIN BROWNLOW

Wings

The aviators constituted the most colorful group of veterans. The postwar slump forced many of them back to flying in the hope of making a living. Provided they could raise from $300 to $600, they could purchase a war-surplus plane, boxed-up and brand-new.

Congress had made a huge appropriation in 1917 for aircraft production. Yet a Universal industrial film of 1918, *The Yanks Are Coming*, shot at the Dayton-Wright Factory, which shouted the praises of the aircraft industry, was suppressed by George Creel [head of the U.S. official propaganda organization, the Committee on Public Information]. Charges of "aircraft lies" hurled at the CPI turned out to be all too true. The House Report of the 66th Congress stated that although more than $1 billion had been spent on combat planes, not a single aircraft reached the battle zone. With the lifting of the wartime ban on civilian flying, former military pilots applied to purchase the surplus aircraft. Enormous quantities were available—but they were mostly trainers. The combat aircraft were scrapped to keep them off the market.

The pilots, known as barnstormers, earned their living doing stunts at county fairs and taking passengers at a dollar a minute. To attract customers to the airfield they staged stunts. The movie studios hired the more skillful for serials and melodramas. Captain Nungesser, the French ace, who had shot down a reported one hundred and five German planes, was signed by the Arcadia Pictures Company of Philadelphia. T. Hayes Hunter directed him in *The Sky Raider* (1925). Nungesser was lost in an attempted transatlantic flight in 1927. Thomas Ince encouraged civil aviation with a $50,000 prize for the first flight across the Pacific. But the most encouraging news of all was the announcement in 1926 of Paramount's massive epic of war in the air, *Wings*.

Irvin Willat's *Zeppelin's Last Raid* (1917) was on of the few pictures about wartime aviation, and most of the exteriors were achieved with miniatures. In 1918 *The Romance of the Air* featured Lieutenant Bert Hall, one of the only two surviving Americans of the original Lafayette Escadrille. The other, Major William Thaw, declined to appear in the film, and his part was played by Herbert Standing. Adapted by Franklin B. Coates from Bert Hall's novel *En l'air*, it was directed by Harry Revier and received government cooperation in the form of combat planes, observation balloons, and hydroplanes. This material was intercut with footage filmed by Lieutenant Hall's observer, including the destruction of a German biplane and the blowing up of an observation balloon. *Moving Picture World* considered the film little more than a publicity vehicle for Hall: "Lieutenant Hall rings true, but his story does not."

Wings was the brainchild of John Monk Saunders, who had been a combat pilot at nineteen. Paramount was anxious for a war epic of its own, and, in February 1926, Jesse Lasky met Saunders in a New York hotel. Lasky was infected by Saunders's enthusiasm. "Go to Washington," he said. "Talk to the government. If they will help us with the picture, we will make it."

Official approval was eventually granted, subject to several conditions: Paramount must pay for all damage done in making the picture, must carry $10,000 insurance for each man who worked on it, must agree not to release it until it had been approved by the War Department, and must keep the fact of the government's cooperation out of the papers as long as possible, so that foreign countries would not think it was being made to intimidate them. Furthermore, the film was to be made in such a manner that it was training for those concerned.

Paramount agreed; the story was handed over to scenario writers Louis D. Lighton and Hope Loring. It became a West Coast production, which meant that B. P. Schulberg was in charge. When Schulberg had joined Paramount, he brought with him Clara Bow. She was given a leading role, along with former aviator Richard Arlen, Charles "Buddy" Rogers, and Jobyna Ralston. A small part was assigned to a young man from a Montana ranch, Gary Cooper. In giving the direction to William Wellman, Schulberg was taking a risk. But although relatively new to direction, Wellman had made a commercial and critical success, *You Never Know Women*. And he had one advantage over the other available directors — he had been an aviator in the war [in the Lafayette Flying Corps.]. Wellman, before he took on this monumental task, watched *The Big Parade* again and again, just as Howard Hughes, before he made *Hell's Angels*, the subsequent aviation epic, ran *Wings* ceaselessly.*

*King Vidor says Wellman gave him one of the ideas for *Big Parade*—the episode where a German plane flies down over a column of troops and its wheels skim the helmets. Wellman himself had done this stunt during the war.

Harry Perry, who had made several pictures for Schulberg in his independent days and had worked on *The Vanishing American* with the supervisor of *Wings*, Lucien Hubbard, took control of the cinematography. He was given several weeks to make tests. The usual method for shooting aerial scenes was to establish the planes in the air, then cut to close-ups shot on the ground, against the sky. But Wellman stipulated there was to be no faking; close-ups of pilots in the air had to be shot in the air. Assisting Perry on the tests were Captain (later General) Bill Taylor and Lieutenant Commander Harry Reynolds. They acquired a gun-ring, a scarf mount for a machine gun, and secured it to the cockpit of the camera plane. A camera mount was made, giving the operator complete freedom. The problem of filming the actors was more difficult.

"We had cameras shooting both forward and backward," said Lucien Hubbard. "There were many scenes where we shot over Arlen's head, as he dived to earth. You put that much weight in an unaccustomed place, and you're courting a crash. You're doing something that has never been accomplished before, and to do it so you don't crash and still get your picture, is really amazing."

The cameras were attached to the aircraft on mounts made in the shape of a saddle, secured with two straps around the body of the plane, and situated in the rear cockpit. This mount was used mostly for motor-driven cameras, shooting forward over the head of the real pilot, showing the actor apparently in a single-seater.

One trick aroused widespread comment, and for a while Paramount refused to admit how it was done (the front office probably didn't know). Frank Clarke, playing a German, was instructed to go up to six thousand feet, then come swirling down, his plane on fire and out of control. This was not so hard to do in a long shot. What astounded audiences — and technicians — was the fact that the camera stayed on him in close-up, the clouds wheeling round behind him. Clarke was equipped with a button-operating Mitchell camera, run by motor, using three 45-volt dry-cell batteries. He reached the required height, switched on the camera, jerked as though hit by a bullet, opened his mouth to release a gush of theatrical blood, released a gate in a box containing lamp-black, let go of the stick, and kicked the plane into a tailspin with his foot.

Aviators acknowledged this to be one of the toughest stunts on record — to sit limp and useless while your plane does the one thing you have been trained to avoid.

For scenes involving actors who could not fly, a large headrest obscured the pilot in the rear cockpit. This automation meant that the exposure had to be adjusted on the ground, the result of guesswork and experience with previous attempts. Filters were dispensed with so lenses could work at the smallest aperture, thus carrying background focus. Harry Perry remembered that after two successive takes of a staff car being attacked, a scene involving hundreds of troops, a fly was found splashed across the center of the lens.

Perry was in command of an army of cameramen. One of the most important of these was Akeley specialist E. Burton Steene, former newsreel man who had served in Mexico, and in 1921 filmed the aerial scenes for a German picture about the country's wartime air supremacy. Steene had been flying since 1912 but regarded *Wings* as the greatest thrill of his life.

In addition to the Akeley, Steene had three Eyemos. Harry Perry, Faxon Dean, and Paul Perry filmed from another group of planes at 1,200 feet. The pilot, Captain Stribling, used a thin rope to signal to Steene when to start cranking. A dud was dropped to get the range. Then came two tugs on the rope.

"It was a wonderful sight to see these death-dealing messengers speeding down — the terrible explosions took place right on schedule, due to the unerring eye and hand of Captain Stribling. For several seconds the ship shook and trembled with each explosion until I thought it might be out of control. The sensation of being rocked and thrown about in a giant bomber a scant 600 feet above the ground while dropping 1,200 pounds of TNT is a thrill not often given to a man. In my cramped quarters it would have been very difficult if not impossible to get away with my parachute, but my confidence in the pilot kept me in repose."

"There was a fatal accident," said Lucien Hubbard. "Some planes came over a hill and dove on the troops. Wellman and I were watching from a platform down below it. Suddenly he turned away and said, 'Oh, my God, no!' And I wondered what was the matter. He had seen a plane come down at such an angle it couldn't possibly go up again, and it crashed right behind us over the ridge. The boy was killed. We thought they would stop the picture, but the operations officer at Kelly Field said, 'What's the matter? What are you so glum about? We don't think anything of it. Guys are always killed in training. It was his own fault. We told him to come in at a certain angle, he came in too short. These things happen. It'll never be heard of.' It never was."

The cooperation of the U.S. government tied up thousands of soldiers, virtually all the pursuit planes the air force had, billions of dollars' worth of equipment — and some of the finest military pilots in the country. Several of those who flew for *Wings* became generals in World War II: Hoyt Vanderberg, Frank Andrews, Hal George, Earl E. Partridge, Clarence Irvine, Rod Rogers, Bill Taylor. The company had the use of the Air Service Ground School at Brooks Field and the advanced flying school at Kelly Field, both near San Antonio, and fliers came from as far away as San Francisco and Virginia.

The U.S. Army's 2nd Division, which had taken part in *The Big Parade*, was used for the big drive at St. Mihiel. This division was the same one that had participated in the original drive. The spot chosen for the location outside San Antonio was a section of ground about a mile long. The grass was so tall that Wellman climbed onto the shoulders of one of the crew to look out across the top of it. In a short time, the place was bombarded by field guns with live rounds — giving the gunners some useful practice — and a huge number of Mexican laborers dug trenches to strict military specifications. The effect was so alarmingly evocative that Wellman, flying over it for the first time, felt a wave of sickness and muttered, "God — I'm back in France."

The location was selected with aerial photography in mind, but when Perry made some tests, he found that at normal flying speed, the plane swooped over it too fast to capture detail. So a hundred-foot tower was erected, with platforms every twenty feet. It was from this huge tower that the main battery of cameras photographed the drive — although Wellman positioned hand-held Eyemos in other strategic positions.

In Washington, angry speeches were made about the frivolous squandering of military resources. The army, which had initially welcomed this chance of recreating warfare, grew sulky as scenes were shot over and over again. Troops grumbled that the aviators were stealing the picture.

The major battle was planned as carefully as a regular military operation; it was realized that with the intensity of explosions, men's lives were actually at stake. After the day's shooting, generals, officers, and engineers held a conference with Wellman and Hubbard, and with blueprints and field maps, they planned the following day, making out a complete battle order for defense, attack, hour of advance, distance, troops, and equipment. This order went to under-officers, who relayed it to squad leaders. These appointed the "killed" and directed the survivors.

"When the director was not working on the picture," said *Motion Picture Magazine*, "he and the producer were settling the difficulties that arose because of army politics, smoothing over jealousies between different branches of the service, pacifying the impatient troops with barrels of beer and motion picture entertainments, giving the aviators dinners and dances and conciliating fuming officers with a diplomacy that would have avoided the World War. The Government cannot be hired, but at the close of the picture, Wellman handed over fifty thousand dollars as a gift to the mess funds."

At the preview in San Antonio, *Wings* ran fourteen reels. The town went wild over it. Nevertheless, Paramount withdrew it, cut it ot 12,267 feet, and premiered it in New York on August 12, 1927, with the additional attraction of Magnascope — a process in which the big scenes were projected onto a greatly enlarged screen. Some authorities declare a General Electric sound-effects track was heard at the premiere, but this seems to have been added the following year — effects machines duplicated the sound of machine guns and airplane motors until the actual sound-effects track was available.

Reviewers, whatever they thought of the rest of the story, were astounded by the aerial and action sequences. Technical faults were few, and the picture was regarded as generally satisfactory even to the aviator; the main criticism was that the Gotha bomber, in actuality a Martin MB4, flew in at a suicidal height — but otherwise it could not have been shown in conjunction with the village.

Moving Picture World interviewed veteran fliers, and invited them to submit a list of errors. Donald Hudson (D.S.C., Croix de Guerre, 27th Aero Squadron, Ist Pursuit Group, AEF) found a few faults, but declared that the action was so convincing that faults did not destroy the illusion of actual combat. "This was clearly demonstrated when the man sitting next to me, who saw several months of service at the front as a pilot, audibly called out 'maneuver' as the pilot on the screen failed to do so when attacked. The crashes are apparently real crashes, and it makes your blood run cold to see them."

Charles Porter (D.S.C., with Oak Leaf Cluster, Croix de Guerre, 147th Aero Squadron, Ist Pursuit Group, AEF) had been frankly skeptical about any moving pictures of aerial combat. "I had seen the official pictures of the Air Service taken in France by the U.S. Singal Corps. These pictures were not only uninteresting, but also unsatisfactory despite the best efforts of experienced photographers. After viewing *Wings*, however, I know that no one need miss seeing the actual manner in which our Air Force conducted its warfare in France."

Porter considered that the detail was so accurate that only someone familiar with the aerodromes in the zone of advance could find the slightest flaw. He pointed out that when a plane was shot down, it seldom burst into flames. He felt that Arlen, when attacked by Rogers, should have taken evasive action; "he could easily have spun or sideslipped to the ground and escaped with a forced landing." All fliers objected to the ending of the picture, which they found "weak and improbable."

Wings gets better the older it becomes. Values that once seemed overly sentimental now seem so much a part of their time that they no longer irritate. For most of its length, *Wings* avoids the grimness of war and captures exactly the fierce romanticism that so many veterans feel for it. But at the end Buddy Rogers shoots down his friend (Richard Arlen), escaping from the German lines in a German plane. Buddy Rogers lands in front of a war cemetery and steps out of his plane, in front of thousands of white crosses.

Otherwise, it is "the road to glory" and "the bravest of the brave" that figure in the titles, and the boys' initial hatred for each other is dispelled by a boxing match in which David is beaten to a pulp but refuses to admit defeat.

Wellman hurls his camera around the vast battlefield wtih exhilarating abandon. Even by today's standards, his setups seem remarkably bold. Although the troops die with rather operatic gestures, his epic handling of the big drive is overwhelming, and the superimposition of thousands of men marching into a horizon where their destruction is pictured in split screen is a moment worthy of Abel Gance's classic *J'accuse*.

The film is dedicated to Charles Lindbergh, and the dedicatory title was adapted from one of his speeches: "To those young warriors of the sky, whose wings are forever folded about them, this picture is reverently dedicated."

The road to Kitty Hawk was paved with the efforts of many a fantasist and scientist, but between Leonardo and the Wright brothers, the man who most advanced the theory of flight was Sir George Cayley. In the period 1796–1809 he flew models of a helicopter and a glider and designed a full-sized modern glider as well as a fixed-wing airplane with tail-unit control. In later years he designed the first biplane, sent up a triplane glider with a boy in the underslung car, distinguished between lift and drag, and calculated the power required to fly airplanes of differing speeds and loads. The seminal paper excerpted below, "On Aerial Navigation," was published in 1809 and is universally acknowledged as the basis of modern aerodynamics.

GEORGE CAYLEY

On Aerial Navigation

The scheme of flying by artificial wings has been much ridiculed, and indeed the idea of attaching wings to the arms of a man is ridiculous enough. . . . The flight of a strong man by great muscular exertion, though a curious and interesting circumstance . . . would be of little use. I feel perfectly confident, however, that this noble art will soon be brought home to man's general convenience, and that we shall be able to transport ourselves and families, and their goods and chattels, more securely by air than by water, and with a velocity of from twenty to one hundred miles per hour. To produce this effect it is only necessary to have a first mover, which will generate more power in a given time, in proportion to its weight, than the animal system of muscles.

. . . it is perfectly clear that force can be obtained by a much lighter apparatus than muscles. . . . we may be justified in the remark that the act of flying . . . requires less exertions than from the appearance is supposed. . . . the many hundreds of miles of continued flight exerted by birds of passage . . . [and] the perfect ease with which some birds are suspended in long horizontal flights, without one waft of the wings, encourages the idea that a slight power only is necessary. . . .

All [the] principles upon which the support, steadiness, elevation, depression, and steerage of vessels for aerial navigation depend have been abundantly verified by experiments both upon a large and small scale. Last year I made a machine having a surface of 300 square feet, which was accidently broken before there was an opportunity of trying the effect of the propelling apparatus, but its steerage and steadiness were perfectly proved, and it would sail obliquely downward in any direction according to the set of the rudder. . . . Even in this state, when any person ran forward in it with his full speed, taking advantage of a gentle breeze in front, it would bear upward so strongly as scarcely to allow him to touch the ground, and would frequently lift him up and convey him several yards together.

The best mode of producing the propelling power is the only thing that remains yet untried toward the completion of the invention. . . .

In combining the general principles of aerial navigation for the practice of the art, many mechanical difficulties present themselves which require a considerable course of skillfully applied experiments, before they can be overcome. But to a certain extent, the air has already been made navigable, and no one . . . can doubt the ultimate accomplishment of this object. . . .

When Wilbur Wright was "about to begin a systematic study" of mechanical and human flight in 1899, he wrote to the Smithsonian Institution inquiring what he might profitably read "to avail [himself] of all that is already known." The reply suggested that one of the books he might try was Octave Chanute's *Progress in Flying Machines*, published in 1894. As it developed, no book had greater influence on the Wright brothers, and no man did more to hone their observations and deductions than Chanute, with whom the Wrights corresponded and exchanged ideas for several years before their first powered flight and after. The passage below forms the conclusion of Chanute's epochal work.

OCTAVE CHANUTE

Progress in Flying Machines

To the possible inquiry as to the probable character of a successful flying machine, the writer would answer that in his judgment two types of such machines may eventually be evolved; one, which may be termed the soaring type, and which will carry but a single operator, and another, likely to be developed somewhat later, which may be termed the journeying type, to carry several passengers, and to be provided with a motor.

The soaring type may or may not be provided with a motor of its own. If it has one this must be a very simple machine, probably capable of exerting power for a short time only, in order to meet emergencies, particularly in starting up and in alighting. For most of the time this type will have to rely upon the power of the wind, just as the soaring birds do, and whoever has observed such birds will appreciate how continuously they can remain in the air with no viable exertion. The utility of artificial machines availing of the same mechanical principles as the soaring birds will principally be confined to those regions in which the wind blows with such regularity, such force, and such frequency as to allow of almost daily use. . . ."

This is the type of machine which experimenters with soaring devices heretofore mentioned have been endeavoring to work out. If unprovided with a motor, an apparatus for one man need not weigh more than 40 or 50 pounds, nor cost more than twice as much as a first-class bicycle. Such machines therefore are likely to serve for sport and for reaching otherwise inaccessible places, rather than as a means of regular travel, although it is not impossible that in trade-wind latitudes extended journeys and explorations may be accomplished with them; but if we are to judge by the performance of the soaring birds, the average speeds are not likely to be more than 20 to 30 miles per hour.

The other, or journeying type of flying machines, must invariably be provided with a powerful and light motor, but they will also utilize the wind at times. They will probably be as small as the character of the intended journey will admit of, for inasmuch as the weights will increase as the cube of the dimensions, while the sustaining power only grows as the square of those dimensions, the larger the machine the greater the difficulties of light construction and of safe operation. It seems probable, therefore, that such machines will seldom be built to carry more than from three to ten passengers, and will never compete for heavy freights, for the useful weights, those carried in addition to the weight of the machine itself, will be very small in proportion to the power required. Thus M. *Maxim* provides his colossal airplane (5,500 square feet of surface) with 300 horsepower, and he hopes that it will sustain an aggregate of 7 tons, about one-half of which consists in its own dead weight, while the same horsepower, applied to existing modes of transportation, would easily impel — at lesser speed, it is true — from 350 to 700 tons of weight either by rail or by water.

Although it by no means follows that the aggregate cost of transportation through the air will be in proportion to the power required, the latter being but a portion of the expense, it does not now seem probable that flying machines will ever compete economically with existing modes of transportaion. It is premature, in advance of any positive success, to speculate upon the possible commercial uses and value of such a novel mode of transit, but we can already discern that its utility will spring from its possible high speeds, and from its giving access to otherwise unreachable points.

It seems to the writer quite certain that flying machines can never carry even light and valuable freights at anything like the present rates of water or land transportation, so that those who may apprehend that such machines will, when successful, abolish frontiers and tariffs are probably mistaken. Neither are passengers likely to be carried with the cheapness and regularity of railways, for although the wind may be utilized at times and thus reduce the cost, it will introduce uncertainty....

The actual speeds through the air will probably be great. It seems not unreasonable to expect that they will be 40 to 60 miles per hour soon after success is accomplished with machines provided with motors, and eventually perhaps from 100 to 150 miles per hour. Almost every element of the problem seems to favor high speeds, and, as repeatedly pointed out, high speeds will be (within certain limits) more economical than moderate speeds. This will eventually afford an extended range of journey — not at first probably, because of the limited amount of specially prepared fuel which can be carried, but later on if the weight of motors is still further reduced. Of course in civilized regions that supply of fuel can easily be replenished, but in crossing seas or in explorations there will be no such resource.

It seems difficult, therefore, to forecast in advance the commercial results of a successful evolution of a flying machine. Nor is this necessary; for we may be sure that such an untrammeled mode of transit will develop a usefulness of its own, differing from and supplementing the existing modes of transportation. It certainly must advance civilization in many ways, through the resulting access to all portions of the earth, and through the rapid communications which it will afford.

It has been suggested that the first practical application of a successful flying machine would be to the art of war, and this is possibly true; but the results may be far different from those which are generally conjectured. In the opinion of the writer such machines are not likely to prove efficient in attacks upon hostile ships and fortifications. They cannot be relied upon to drop explosives with any accuracy, because the speed will be too great for effective aim when the exact distance and height from the object to be hit cannot be accurately known. Any one who may have attempted to shoot at a mark from a rapidly moving railway train will probably appreciate how uncertain the shot must be.

For reconnoitering the enemy's positions and for quickly conveying information such machines will undoubtedly be of great use, but they will be very vulnerable when attacked with similar machines, and when injured they may quickly crash down to disaster. There is little question, however, that they may add greatly to the horrors of battle by the promiscuous dropping of explosives from overhead, although their limited capacity to carry weight will not enable them to take up a large quantity, nor to employ any heavy guns with which to secure better aim.

Upon the whole, the writer is glad to believe that when man succeeds in flying through the air the ultimate effect will be to diminish greatly the frequency of wars and to substitute some more rational methods of settling international misunderstandings. This may come to pass not only because of the additional horrors which will result in battle, but because no part of the field will be safe, no matter how distant from the actual scene of conflict. The effect must be to produce great uncertainty as to the results of maneuvers or of superior forces, by the removal of that comparative immunity from danger which is necessary to enable the commanding officers to carry out their plans, for a chance explosive dropped from a flying machine may destroy the chiefs, disorganize the plans, and bring confusion to the stronger or more skillfully led side. This uncertainty as to results must render nations and authorities still more unwilling to enter into contests than they are now, and perhaps in time make wars of extremely rare occurrence.

So may it be; let us hope that the advent of a successful flying machine, now only dimly foreseen and nevertheless thought to be possible, will bring nothing but good into the world; that it shall abridge distance, make all parts of the globe accessible, bring men into closer relation with each other, advance civilization, and hasten the promised era in which there shall be nothing but peace and goodwill among all men.

This small, perfect passage describes an early peril in Sir Francis Chichester's *Ride on the Wind*, the account of his 1931 attempt at a solo seaplane flight around the world. The flight, begun in Sydney, Australia, ended in disaster at Katsuura, Japan, when erroneous instructions from the harbormaster caused Chichester to fly into some telegraph wires which catapulted him into a harbor wall; his plane, the *Gypsy Moth*, was reduced to rubble and so, very nearly, was he: thirteen broken bones and five years spent as an invalid. But Chichester recovered to resume his adventuring ways, eventually attaining knighthood and the pinnacle of world fame in 1967 when, at age sixty-five, he sailed solo around the world. He named his yacht for the seaplane that had crashed so ignominiously three decades before.

FRANCIS CHICHESTER

A Wheelbarrow With Wings

The word Merauke should mean mud. The flat, flowing river was muddy, the banks were mud; the low-lying country all round put one in mind of sloppy mud, as did also the reeds, the willowlike scrub and the swamp grass.

I taxied downriver to the far side and headed back into the slight easterly breeze. The air was hot and sticky, the water surface tired. The plane failed to rise. I turned and it seemed to take an age to taxi a mile or perhaps a mile and a half downriver again. And it was no good, although I held on until nearly into the bank at the river bend. I stopped the motor and let it cool while the plane drifted slowly. The heat of the motor met my face. I know nothing that settles despair to the depths of one's heart so much as failing to take off. Why it should do so, I cannot conceive. My project seemed a terribly futile thing, trying to fly round the world in a ridiculous little seaplane, totally inadequate. I drifted in dreary silence, with the heat irritating and fitting me like clothes of hair, and later I taxied further down till I was over by the reeds straggling in the muddy water.

I must have had two miles of clear river ahead of me then. The seaplane seemed to climb out of the water till on the step, when by all the rules she ought to have come away with ease; but as soon as I tried to stroke her off I could feel the suction of the smooth water pull at the float heels and slow her down. Then, time after time, I pig-jumped her off with a jerk, but she would not keep in the air and settled back. Yet she was nearly off. But now the bank in front was rushing up fast. At the previous run I had noticed how very nearly the seaplane was to taking off close to the bank.*

I had no right to go another inch, and I knew it; if I pig-jumped her once more and she failed to keep the air, it was a crash. Some desperate feeling drives one. I held her down to the last few yards, then pulled her off sharply. She jumped the bank. I had been concentrated on the bank; I looked up to find a clump of low trees filling the space in front and to the left. Obviously I could not get over them. I could not turn to the left where the river ran before the trees. I could force the seaplane down to the mud and willowy scrub beneath me, with probable safety to myself but certain destruction to the plane. I could not do that. I was already in a skidding turn to the right. Small-leaved shrubs like willows came at me; I had not clearance enough above them to bank properly, and the seaplane skidded toward the trees. If only the plane did not feel so infernally heavy in the air. The slots were hanging out; the plane had scarcely flying speed. I watched the willow tops under the right wing tip, banking as much as I dared without the wing tip touching. I was aware of the willows under the float; it must touch, but so long as no stump in the small branches caught it. . . . It was the skid I feared. Underbanked, she was skidding to the left. The trees were there. I could not look up from the willows under the right wing tip. I felt as if I were an animal looking away while waiting to be shot. What ages I waited while life stood still!

No crash. I must be round. . . . Coconut trees ahead. I jumped above their tops and skimmed them, gathering speed. It was like flying a mud-clogged old wheelbarrow with wings. Before I left that sea of tufted coconut tops I suddenly felt sick to death of them.

*It may have been shallow there and the float-heels picking up the recoil of the bow-waves from the bottom.

Now that space is no longer the sole province of Buck Rogers and his ilk, yesterday's lunatic has become today's seer. One such prophet who had to wait to find honor in his own country is Arthur C. Clarke, an original member of the British Interplanetary Society of the 1930s and author of the more recently prophetic *2001* and *2010*. Somehow, as he observes in those novels and in this article written in 1963, the fun has gone out of contemplating man's future in space.

ARTHUR C. CLARKE

Armchair Astronauts

Today we regard the first landing on the moon as inevitable, and there will be harsh words in Congress and the Kremlin if it occurs on the wrong side of 1970. Yet to some of us, the heroic period of the Space Age lay between 1935 and 1955, and what has happened since has a slight air of anticlimax. In those days, we would-be astronauts were a small band of pioneers, surrounded on all sides by sniping scientists who denounced us as irresponsible crackpots. Now the only sort of argument one hears about space is whether Brobdingnag Astrodynamics or Consolidated Aerospace should be awarded the $326,709,163 contract for the first-stage Mastodon booster.

It was very different in the prewar years. Then the annual income of the British Interplanetary Society was about $300; as treasurer, I had the terrifying responsibility of accounting for it.

On the other side of the Atlantic, the American Rocket Society was slightly more affluent, but as we both operated with a part-time, voluntary secretarial staff, contact between the organizations was erratic. In those days, moreover, the B.I.S. and the A.R.S. were divided by an ideological gulf.

As is well known, we British are a romantic and wildly imaginative people, and to our annoyance, the conservative Americans did not consider space travel respectable. Though they had formed the American *Interplanetary* Society in 1930, the name had been changed to American *Rocket* Society a few years later. The suggestion was sometimes made that we should follow suit, but we refused to lower our sights. To us the rocket was merely the interplanetary bus; if a better one came along (it hasn't yet, but we're still hoping) we would transfer and give the rocket back to the fireworks industry.

Picture us, then, in the mid-1930s, when only a few aircraft had flown at the staggering speed of three hundred miles an hour, trying to convince a skeptical world that men would one day travel to the moon. There were about ten of us in the hard core of the society, and we met at least once a week in London cafés, pubs or one another's modest apartments.

We were almost all in our twenties, and our occupations ranged from aeronautical engineer to civil servant, from university student to stock-exchange clerk. Few of us had a technical or scientific education, but what we lacked in knowledge we made up in imagination and enthusiasm. It was, I might add, just as well that we were overoptimistic. If we had even dreamed that the first round trip to the moon would cost $10 billion per passenger, and that spaceships would cost many times their weight in gold, we should have been much too discouraged to continue our quarter-million-mile uphill struggle.

The total amount spent on the British space effort before the outbreak of World War II was less than $1,000. What did we do with all that money? Most of us talked, some of us calculated and a few of us drew — all to considerable effect. Slowly there emerged the concept of a space vehicle that could carry three men to the moon and bring them back to earth. It had a number of unconventional features even for a 1938 spaceship. Most of them are commonplace today, and many have been "rediscovered" by later workers. Notable was the assumed use of solid propellants, of the type now employed in Polaris and similar missiles.

Our first plans, based on highly unrealistic assumptions, envisaged making the entire round trip in a single vehicle, whose initial weight we hopefully calculated at about a thousand tons (the Advanced Saturn now being developed by NASA will weigh three times as much). Later we discussed many types of rendezvous and space-refueling techniques, hoping to break down the journey into manageable stages. One of these involved the use of a specialized "ferry" craft to make the actual lunar landing while the main vehicle remained in orbit. This, of course, is the approach now being used in the Apollo Project — and I am a little tired of hearing it described as a new idea. For that matter, I doubt if we thought of it first; it is more than likely that the German or Russian theoreticians had worked it out years before.

There is a vast gulf, almost unimaginable to the layman, between thinking up a plan and converting it into detailed engineering blueprints. There is an equally great gulf between the blueprints and the final hardware, so we cannot claim too much credit for our pioneering insight. Yet I am often struck by the fact that there is hardly a single new conception in the whole field of current space research; most of what is happening now was described, at least in outline, twenty or even fifty years ago.

But back to our Model T. As soon as we had finished the drawings, we published them in our minute *Journal of the British Interplanetary Society.* It took us some time to collect enough money to pay the printer—he was a Greek, I remember, and a few Hellenic spellings slipped through my proofreading. Nor am I likely to forget the day I collected the entire edition, in two parcels, and was walking home with it to the apartment I shared with another space enthusiast a few blocks east of the British Museum. I was halfway home when two polite gentlemen in mackintoshes tapped me on the shoulder and said, "Excuse me, sir, but we're from Scotland Yard. Could we see what you have in those packets?"

It was a reasonable request, for at the time wild Irishmen were blowing up post offices to draw attention to their grievances, and the Yard was trying to round them up. (They did catch a fellow named Brendan Behan, I believe.) To the considerable disappointment of the detectives, I was not even carrying *Tropic of Capricorn,* but when I presented them with copies of the *Journal,* they very gamely offered to pay. Tempting though it was to acquire a genuine subscriber (the cashbox held about $2.50 at the time), I refused the contribution. But I did get them to carry the parcels the rest of the way for me.

The *Journal* attracted a surprising amount of attention and a not surprising amount of amusement. That *doyen* of scientific publications, the good gray *Nature,* condescended to notice our existence, but concluded its review with the unkind cut: "While the ratio of theorizing to practical experimentation is so high, little attention will be paid to the activities of the British Interplanetary Society."

That was a quite understandable comment, but what could we do about it with only $2.50 in the till? Why, we could launch an appeal for an Experimental Fund.

We did so, and the money came rolling in. There was one occasion, I now blush to recall, when I shared sardines on toast with an elderly lady member in an Oxford Street tearoom and convinced her that for sixty dollars one could solve the basic problems of building a meteorological rocket. Eventually we rounded up a couple of hundred bucks, and the research program was under way. At Peenemünde, though we were not to know it for quite a while, von Braun was already heading for his first hundred million.

All this money was something of a responsibility; having appealed for it, we had to use it in a manner most calculated to produce both scientific results and publicity. The actual building and launching of rockets was frowned upon by the officials, who would certainly start police proceedings under the 1875 Explosives Act—as the efforts of a group of experimenters in the north country had already proved.

We were in the position of someone who couldn't afford a car but had enough for the speedometer and the rearview mirror. This analogy is quite exact; though we couldn't make a down payment on even a compact spaceship, we felt we could develop two of the instruments needed to operate it. It was a sensible decision, indeed about the only one possible in the circumstances.

The first project we tackled was the construction of a spaceship speedometer which had been invented by Jack Edwards, the eccentric genius who headed our research effort. Edwards, who is now dead, was a short, bearded and excitable Welshman, and the nearest thing to a mad scientist I have ever met outside fiction. He was the director of a very small electronics firm which soon afterward expired, thanks to his assistance; but he had an altogether uncanny grasp of the principles of astronautics. Back in 1938, he had invented what is now called inertial guidance — the technique that allows a rocket to know just where it is and how fast it is going by continually keeping track of the accelerations acting upon it.

Edwards's space speedometer consisted of a large aluminum disc pivoted on ball bearings and with sundry gears, weights and springs attached. As the device was moved up or down, the weights would "sense" the forces acting upon them, and the rotation of the disc would record the distance moved. We had planned to test the gadget on one of the deeper elevators of the London Underground, but we never got as far as that. The theory of the device was perfectly sound, and something similar steers every satellite into orbit today. But the engineering precision demanded was utterly beyond our means, and Mrs. Edwards put her foot down on hearing of our intention to cast lead weights in her best saucepan.

Balked on the speedometer front, we tried our luck with the rearview mirror. To keep our spaceship on course during takeoff and to provide the crew with artificial gravity, we had proposed to spin the whole thing like a rifle bullet. The spin would be imparted by water jets while the ship floated in a vertical position before launching. (The prospective launching site we favored was a raft on Lake Titicaca, high up in the Andes.) Even though the rate of rotation was quite low, it would obviously be impossible to take observations of the stars from our cosmic carousel, so we had to invent an optical system to unscramble the ship's spin.

This required no great originality, for the astronomers — who also look out at the stars from a spinning vehicle, the planet Earth — had solved the problem years before with an instrument called a coelostat, However, it only has to cope with one revolution every twenty-four hours. We built a similar arrangement of four mirrors — two fixed, two spinning — and I sacrificed the spring motor of my phonograph to provide the motive power.

The coelostat worked. It was the only thing we ever made that did. Its public demonstration took place in most auspicious surroundings — the hallowed halls of the South Kensington Science Museum, whose director deserves much credit for providing hospitality to such a far-out organization as ours.

Our setup was simple but effective. At one side of the room was a disc with lettering on it, spinning too rapidly for the words to be read. At the other was the coelostat — a wooden box measuring about a foot along each side, looking rather like the result of a *mésalliance* between a periscope and an alarm clock.

When you peered through the coelostat at the spinning disc, the latter seemed stationary, and you could read the inscription "BIS" painted on it. If you looked at the rest of the room through it, everything appeared to be revolving rapidly; this was not recommended for any length of time.

Though our experimental efforts were unimpressive, we made ourselves known through countless lectures, newspaper interviews and argumentative letters to any publications that would grant us hospitality. One controversy ran for months in the correspondence columns of the BBC's weekly, *The Listener.* If we could not convince our critics, we usually routed them.

Looking back on it, I am amazed at the half-baked logic that was used to attack the idea of space flight. Even scientists who should have known better employed completely fallacious arguments to dispose of us. They were so certain that we were talking nonsense that they couldn't be bothered to waste sound criticism on our ideas.

My favorite example of this is a paper which an eminent chemist presented to the British Association for the Advancement of Science. He calculated the energy a rocket would need to escape from the earth, made a schoolboy howler in the second line and concluded, "Hence the proposition appears to be basically impossible." But that was not enough, he could not resist adding, "This foolish idea of shooting at the moon is an example of the absurd lengths to which vicious specialization will carry scientists working in thought-tight compartments." I cannot help feeling that the good professor's compartment was not merely thought-tight; it was thought-proof.

As another example of the sort of stick that was used to beat us, I might mention an article that appeared under the eye-catching title *We Are Prisoners of Fire.* This was based on the fact, deduced from radio measurements, that there are layers in the upper atmosphere where the temperature reaches a couple of thousand degrees Fahrenheit. Therefore, the writer announced smugly, any space vehicle would melt before it got more than a few hundred miles from earth. He had overlooked the point that at those altitudes the air is so thin that the normal concept of temperature has no meaning, and one could freeze to death for all the heat a few 2,000-degree molecules of gas could provide.

I must admit that we thoroughly enjoyed our paper battles; we knew that we were riding the wave of the future. But the world-to-be was moving inexorably, unmistakably, toward war. I remember sending out, from the fourth-floor apartment on Gray's Inn Road that was both my residence and the BIS headquarters, an emotional farewell to all our hundred members, and then descending to the shelters as the sirens gave their warning.

It was a false alarm; nothing happened then, nor for a long time afterward. Finding to our surprise that we had not all been blown to pieces, we resumed contact and continued our discussions by means of correspondence and occasional private meetings. As an RAF instructor, I was in a position to indoctrinate hundreds of hapless airmen, and I made the most of the opportunity. For some odd reason, my service nickname was "Space Ship."

In the winter of 1944, the European conflict was clearly drawing to an end. Nevertheless, for several weeks large holes had been suddenly appearing in southern England, though there was nothing about it in the papers. We were holding a meeting in London to plan our postwar activities. The speaker had just returned from a mission in the United States, where a well-known authority had assured him that tales of large German war rockets were pure propaganda. We were still laughing at this when — CRASH — the building shook slightly, and we heard that curious, unmistakable rumble of an explosion climbing back into the sky — the sound that comes from an object that has arrived faster than the sound of its own passage. A few months later, when we knew his address, we hastened to confer the honorary membership of the Society on Dr. Wernher von Braun.

The post-V-2 world, of course, took us much more seriously. Few people now doubted that rockets could travel great distances into space, and most were prepared to admit that men could travel with them. We altered our propaganda line. It was no longer necessary to spend all our efforts proving that space flight was possible; now we had to demonstrate that it was desirable. Not everyone agreed with us.

One who did was George Bernard Shaw, who joined the Society in his ninety-first year and remained a member until his death. He was a personal capture of whom I was very proud. In 1946, while still at college, I sent him a copy of my Toynbee-inspired, philosophical paper, *The Challenge of the Spaceship*. To my surprise, back came one of the famous pink postcards, followed soon afterward by a longer communication containing some typically Shavian theories of transonic flight. If you are interested, you will find the whole of the brief Shaw-Clarke correspondence in *The Virginia Quarterly Review* for Winter 1960.

Less sympathetic to our aims was Dr. C. S. Lewis, author of two of the very few works of space fiction that can be classed as literature — *Out of the Silent Planet* and *Perelandra*. Both of these fine books contained attacks on scientists in general and astronauts in particular, which aroused my ire. I was especially incensed by a passage in *Prerelandra* referring to "little Interplanetary Societies and Rocketry Clubs" that hoped to spread the crimes of mankind to other planets, and by the words, "The destruction or enslavement of other species in the universe, if such there are, is to these minds a welcome corollary."

An extensive correspondence with Dr. Lewis led to a meeting in a famous Oxford pub, the Eastgate. Seconding me was my friend Val Cleaver, a space buff from way back and today chief engineer of the Rolls-Royce Rocket Division. Needless to say, neither side converted the other. But a fine time was had by all, and when, some hours later, we emerged a little unsteadily from the Eastgate, Dr. Lewis's parting words were, "I'm sure you're very wicked people — but how dull it would be if everyone was good."

The Flying Saucers caused us considerable embarrassment from 1948 on, because there was a danger that in the public eye we should be associated with the cranks and crackpots who were spearheading the cult. In an attempt to strike a blow for sanity, I did a half-hour television program exposing a gentleman who claimed to have made contact with Saucerites. My quest for ammunition led me to a second meeting with Scotland Yard, whose photographic experts examined the crudely faked "evidence" and gave me some useful unofficial advice. I returned to my own darkroom and produced a much better set of flying saucers, proving that any number can play.

Though the Society still had no money, it was a good deal larger than in the prewar days, and the quality of its membership was considerably more impressive. Our bimonthly *Journal* was widely read; in particular, the Soviet embassy subscribed to twenty copies. And there is a very odd thing: though the Russians purchased the *Journal* in bulk and arranged their own distribution, they sent us a complete list of all the scientific and technical institutions in the USSR which received copies. I passed it on to the parties who should have been interested, but as it turned out, they apparently weren't.

By the late 1940s it was obvious that small satellite vehicles could be developed in the near future and would be of enormous scientific value. That they might also be of commercial value I had pointed out as early as 1945, when I first proposed the use of satellites for global radio and television communications. The brilliant success of Telstar has left no doubts about the soundness of the principle, but the system I advocated has yet to be tried in its entirety. It involved the use of three satellites at such an altitude that they would move at the same rate as the turning earth—and so would appear fixed in the sky. NASA's Project Syncom (Synchronous Communications satellite) should be the first step in this direction.

In 1951 all these ideas came to a head when we arranged an international congress in London on the theme of the artificial satellite. It was well attended by scientists from many countries, and one paper described the construction of a satellite vehicle of a size and performance very similar to the later Vanguard. It was designed to put into orbit an inflatable metalized balloon. Less than ten years later the whole world was to watch such an object—the moving star of Echo I.

By this time official circles in the United States had finally started to take a mild interest in space. A few farsighted individuals had already done much more, frequently to the annoyance of their superiors. I once heard General Bernard Shriever, at that time head of the Air Force's ballistic-missile program, remark that he still keeps in his safe a Department of Defense directive forbidding him to use the word "space" in any public statements.

Among the postwar American converts was a young physicist named Fred Singer, then in London as science attaché with the U.S. Office of Naval Research. He had already done notable work with rocket probes in the upper atmosphere, but he was somewhat skeptical about space flight. However, after a few brainwashing sessions he became wildly enthusiastic, and we BIS members soon had to hold him down lest he start galloping all over the solar system. He is now director of the U.S. Weather Bureau's Meteorological Satellite Activities, temporarily assigned to Earth.

One evening Fred, Val Cleaner, and I were sitting in London's celebrated Arts Theatre Club thinking of ways to drum up interest in scientific satellites. "What we want," said someone, "is a nice snappy name for the project." That started us doodling and after a little while we concocted the abbreviation Mouse, for Minimum Orbital Unmanned Satellite of the Earth. In the next few months Fred produced a blizzard of papers describing what Mouse could do: his predictions were uncannily accurate, and every one of them has since come true. The publicity campaign was extremely successful, and Mouse appeared in technical journals all over the world. Indeed, a few years later an American news agency picked up one of Fred's drawings from a Russian paper and hawked it around as one of a genuine Sputnik!

Our conversion of Fred Singer into a space cadet was probably one of the most important things we ever did. He played a dominant role on the committee that recommended the launching of an IGY (International Geophysical Year) satellite. His intervention at a crucial moment was possibly decisive in committing the United States to a satellite program. That is was the wrong satellite was not his fault.

Just as the V-2 marked the end of the first era of astronautics in 1945, so the announcement of Project Vanguard ten years later marked the end of the second. As far as we old space hands were concerned, the long campaign was over. A major power was now in the satellite business, reluctantly but inescapably. Given time, everything we had predicted was bound to follow. Some of us hoped we might live to see the first landing on the moon. In one of my early novels I had stuck my neck out by suggesting 1978 as a target date. Today anyone so pessimistic would be extremely unpopular at NASA headquarters.

There is no need to elaborate upon what has happened since then; it is enough to list Sputnik, Laika, Lunik, Gagarin, Shepard, Titov, Glenn, Mercury, Telstar, Mariner—these are merely the first words in the vast new vocabulary of space. It is a privilege to have watched the beginnings and taken some small part in the great adventure, but now it has grown too huge for the comfort of amateurs like mysefl. This never struck me so strongly as in the fall of 1961, in the Grand Ballroom of New York's Waldorf Astoria.

There, some two thousand scientists and engineers, all in evening dress, had assembled for the banquet that concluded the American Rocket Society's *Space Flight Report to the Nation*. The cream of the astronautics industry, soon to be the largest business in the world, was gathered together. Had the roof fallen in, that would have been the end of the United States' space effort—and of its vice-president, for he was the guest of honor, speaking on a nationwide radio hookup.

It was an impressive occasion, and I was happy to be there. But I could not help thinking of the little pubs and tearooms where we of the B.I.S. dreamed dreams we never thought to see come true. The new generation will know the triumphs, the excitement, the responsibility of space flight. But we had most of the fun.

America's astronauts tend to become celebrities (see Norman Mailer's piece later in this volume), and celebrities often write bad if profitable books. The glorious exception in this arena is Michael Collins's *Carrying the Fire*, from which we have a sweaty-palm account of the rendezvous of the Apollo 11 lunar module (LM) *Eagle*, carrying Buzz Aldrin and Neil Armstrong, with the command module *Columbia* piloted by Collins.

MICHAEL COLLINS

Homeward Bound From the Moon

"*Columbia, Columbia*, good morning from Houston." "Hi Ron." Ron Evans has had the night shift so far this flight, and I haven't talked to him very much. Now he's got the early-morning duty, and he's trying to make sure I'm awake. "Hey, Mike, how's it going this morning?" "How goes it? . . . I don't know yet, how's it going with you?" I'm still groggy. "Real fine here, *Columbia* . . . we're going to keep you a little busy here . . . ," he apologizes. Don't I know it! A space day always seems to start with a bang, with no time even to take a pee before the switch-throwing begins. Today is rendezvous day, and that means a multitude of things to keep me busy, with approximately 850 separate computer key strokes to be made, 850 chances for me to screw it up. Of course, if all goes well with *Eagle*, then it doesn't matter too much, as I merely retain my role as sturdy base-camp operator and let them find me in my constant circle. But if . . . if . . . if any one of a thousand things goes wrong with *Eagle*, then I become the hunter instead of the hunted. Furthermore, the roles can become reversed with no warning at any point along that 850-step path, so I must get up on my tiptoes and stay there, all day long. Neil and Buzz will lift off in slightly over three hours; Ron hasn't awakened them yet. He wants me to get a head start on them by tracking a lunar landmark one last time, to update my computer prior to liftoff. By the time they get up, I have whizzed on by them, and I am halfway through breakfast, fully awake. Things must be stirring in Houston, too, and I imagine there is quite a crowd of people hunched over every available console in Mission Control. Jim Lovell, Neil's backup, comes on the air with some unaccustomed formality. "*Eagle* and *Columbia*, this is the backup crew. Our congratulations for yesterday's performance, and our prayers are with you for the rendezvous. Over." "Thank you, Jim." "Thank you, Jim." Neil and Buzz answer quickly, and I add, "Glad to have a big room full of people looking over our shoulder."

When the instant of liftoff does arrive, I am like a nervous bride. I have been flying for seventeen years, by myself and with others. I have skimmed the Greenland ice cap in December and the Mexican border in August; I have circled the earth forty-four times aboard *Gemini 10*. But I have never sweated out any flight like I am sweating out the LM now. My secret terror for the last six months has been leaving them on the moon and returning to earth alone; now I am within minutes of finding out the truth of the matter. If they fail to rise from the surface, or crash back into it, I am not going to commit suicide; I am coming home, forthwith, but I will be a marked man for life and I know it. Almost better not to have the option I enjoy. Hold it! Buzz is counting down: "9—8—7—6—5— . . . abort stage . . . engine arm ascent . . . proceed . . . beautiful . . . 36 feet per second up. . . ." Off they go: their single engine seems to be doing its thing, the thing earthlings have been insisting it could do for half a dozen years, but it's scary nonetheless. One little hiccup and they are dead men. I hold my breath for the seven minutes it takes them to get into orbit. Their apolune is 47 miles and their perilune is 10 miles. So far so good. Their lower orbit ensures a satisfactory catch-up rate, and they will be joining me in slightly less than three hours, if all goes well.

In the meantime, my hands are full with the arcane, almost black-magical manipulations called for by my solo book. The book is attached by an alligator clip to my helmet tie-down strap, and I am religiously following it line by line, checking off each item—no matter how picayune—as I go. I have locked on to them with my electronic ranging device, which is part of my VHF radio. It reports they are 250 miles behind me, and then promptly breaks lock. I fiddle with it; it reacquires the LM briefly and then breaks lock again. Each time it does this, I must inform Colossus IIA to ignore the data coming from the VHF ranging ("Verb 88, enter") or to start paying attention to it again ("Verb 87, enter"). In between these computer key punches, I keep my eye glued to the sextant which Colossus IIA is pointing at where it thinks the LM ought to be. Sure enough, there is a tiny blinking light in the darkness, and I am able to align my sextant precisely with it and hit the mark button several times. With both VHF ranging and sextant angular data coming in, I am in good shape. Colossus IIA knows precisely where the LM is now, and if for some reason the LM's thrusters act up, I can perform

MAY 5, 1961:
ONE MONTH AFTER THE
U.S.S.R. PUT ITS FIRST
COSMONAUT INTO ORBIT,
AMERICAN ALAN SHEPARD
MAKES A SUBORBITAL
FLIGHT OF 15 MINUTES.

A GIANT LEAP!

FEB. 20, 1962:
JOHN GLENN
BECOMES
THE FIRST
AMERICAN
TO ORBIT
THE EARTH.

MORE SUCCESSFUL
FLIGHTS FOLLOW.

JUNE 3, 1965:
U.S. ASTRONAUT
ED WHITE TAKES
A WALK IN SPACE.

JULY 20, 1969:
NEIL ARMSTRONG IS
THE FIRST MAN TO
SET FOOT ON THE
MOON. AS HE STEPS
DOWN FROM HIS LUNAR
MODULE ARMSTRONG
SAYS: "THAT'S ONE
SMALL STEP FOR A
MAN, ONE GIANT
LEAP FOR MANKIND."

the mirror image of the LM maneuvers and catch it, instead of being caught by it. The continual interruption of data from the VHF ranging is annoying but not serious. We are both on the back side of the moon now, and it is time for the LM to make its first overtaking maneuver, raising itself into a circular orbit some fifteen miles below mine. I am prepared to make the burn if they cannot, and I nervously count them down. "Forty-five seconds to ignition." "OK," says Buzz, and then, "We're burning . . . burn complete, Mike." I punch their numbers into my computer, and it does some calculating and reports our orbits: my apolune is 63.2 nautical miles, my perilune, 56.8. Theirs is now 49.5 by 46.1. In theory, I should be 60 by 60, and they should be 45 by 45, but these numbers are well within acceptable bounds, and everything looks good so far.

There is a strange noise in my headset now, an eerie woo-woo sound. Had I not been warned about it, it would have scared hell out of me. Stafford's Apollo 10 crew had first heard it, during their practice rendezvous around the moon. Alone on the back side, they were more than a little surprised to hear a noise that John Young in the command module and Stafford in the LM each denied making. They gingerly mentioned it in their debriefing sessions, but fortunately the radio technicians (rather than the UFO fans) had a ready explanation for it: it was interference between the LM's and command module's VHF radios. We had heard it yesterday when we turned our VHF radios on after separating our two vehicles, and Neil said that it "sounds like wind whipping around the trees." It stopped as soon as the LM got on the ground, and started up again just a short time ago. A strange noise in a strange place.

Buzz and I are working on a new problem now, measuring any difference in the plane of our orbits. Reassuringly, we both agree that our orbital paths are tilted at precisely the same angle, that we are close enough to perfection that we require no sideways burn at this time to align ourselves more accurately. The LM does make a very small in-plane correction, to reduce its altitude variations as it overtakes me. Shortly thereafter, I pass over the landing site for the first time since they have departed. What a relief! "*Eagle, Columbia* passing over the landing site. It sure is great to look down there and not see you!" Not that I ever did see them on the surface, but to pass over and know they are not stranded down there is worth the price of the entire Apollo program to me.

The LM is fifteen miles below me now, and some fifty miles behind. It is overtaking me at the comfortable rate of 120 feet per second. They are studying me with their radar and I am studying them with my sextant. At precisely the right moment, when I am up above them, 27 degrees above the horizon, they make their move, thrusting toward me. "We're burning," Neil lets me know, and I congratulate him. "That-a-boy!" We are on a collision course now, or at least we are supposed to be; our trajectories are designed to cross 130 degrees of orbital travel later (in other words, slightly over one-third of the way around in our next orbit). I have just passed "over the hill," and the next time the earth pops up into view, I should be parked next to the LM. As we emerge into sunlight on the back side, the LM changes from a blinking light in my sextant to a visible bug, gliding golden and black across the crater fields below. "I see you don't have any landing gear." Of course, only the top half, called the ascent stage, of *Eagle* is returning; the descent stage sits at Tranquility Base for all time; its last (and best) function having been to serve as launch pad. "That's good," chortles Neil. "You're not confused which end to dock with, are you?" Then he adds, "Looks like you are making a high side pass on us, Michael," using fighter-pilot terminology. Buzz sees me, too. "OK, I can see the shape of your vehicle now, Mike." So close, yet so far away: all that remains is for them to brake to a halt using the correct schedule of range vs. range rate. My solo book tells me that at 2,724 feet out, they should be closing at 19.7 feet per second; at 1,370 feet, 9.8 feet per second, etc. While they are doing this, they must make certain they stay exactly on their prescribed approach path, slipping neither left nor right nor up nor down. John Young and I both know that fuel-guzzling whifferdills result if one is not extremely careful, and this is what concerns me now. The sextant is useless this close in, so I close up shop in the lower equipment bay, transfer to the left couch, and wheel *Columbia* around to face the LM.

Goddamn, it looks good! I can look out through my docking reticle and see they are steady as a rock as they drive down the center line of that final approach path. I give them some numbers. "I have 0.7 mile and I got you at 31 feet per second." Buzz replies, "Yes — yes, we're in good shape, Mike; we're braking." Jesus, we really *are* going to carry this thing off! For the first time since I was assigned to this incredible flight six months ago, for the first time I feel that it *is* going to happen. Granted, we are a long way from home, but from here on, it should be all downhill. Bigger and bigger the LM gets in my window, until finally it nearly fills it completely. I haven't touched the controls. Neil is flying in formation with me, and doing it beautifully, with no relative motion between us. I guess he is about 50 feet away, which means the rendezvous is over. "I got the earth coming up . . . it's fantastic!" I shout at Neil and Buzz, and grab for my camera, to get all three actors (earth, moon, and *Eagle*) in the same picture. Too bad *Columbia* will show up only as a window frame, if at all. Within a few seconds Houston joins the conversation, with a tentative little call. "*Eagle* and *Columbia*, Houston standing by." They want to know what the hell is going on, but they don't want to interrupt us if we are in a crucial spot in our final maneuvering. Good heads! However, they needn't worry, and Neil lets them know it. "Roger, we're stationkeeping."

Neil has turned *Eagle* around now, so that its black spot (the drogue used in docking) is directly facing me. Control passes from *Eagle* to *Columbia* at this point, as per our training sessions. It is easier to fly the docking maneuver from the command module; although it can be done from the LM, it is awkward for Neil, because he would have to crane his neck to see out an overhead window, whereas I can look straight ahead in more conventional fashion. So I sight through my reticle and align my probe with *Eagle*'s drogue, just as I did five days ago when I pulled the LM loose from the Saturn. There are a few differences, mainly that the puny little LM ascent stage, nearly empty, weighs less than 6,000 pounds now instead of the nearly 33,000 both stages weigh when they are full of fuel. However, I am not the slightest bit worried as I draw closer and closer. The alignment looks very good indeed at the instant of contact, which I feel as a barely perceptible little nudge.

As soon as we are engaged by the three little capture latches, I flip a switch which fires one of my nitrogen bottles to start the retraction cycle, to pull the two vehicles together. When I do, I get the surprise of my life! Instead of a docile little LM, suddenly I find myself attached to a wildly veering critter·that seems to be trying to escape. Specifically, the LM is yawing around to my right, and we are misaligned by about 15 degrees now. I work with my right hand to swing *Columbia* around, but there is nothing I can do to stop the automatic retraction cycle, which takes some six or eight seconds. All I can hope for is no damage to the equipment so that if this retraction fails, I can release the LM and try again. Things are moving swiftly now, as I wrestle with my right-hand controller. We are veering back toward center line now, and get there, and *bang*, the docking latches slam shut, and miraculously all is well again. Whew! I explain it to Neil and Buzz. "That was a funny one. You know, I didn't feel a shock, and I thought things were pretty steady. I went to RETRACT there, and that's when all hell broke loose." Oops, one never swears on the radio—perhaps that's why I swear so much otherwise. There's no point in worrying about it now. It's time to hustle on down into the tunnel and remove hatch, probe, and drogue, so Neil and Buzz can get through the tunnel. Thank God, all the claptrap works beautifully in this, its final workout. The probe and drogue will stay with the LM and be abandoned with it, for we have no further need for them and don't want them cluttering up the command module. The first one through is Buzz, with a big smile on his face. I grab his head, a hand on each temple, and am about to give him a smooch on the forehead, as a parent might greet an errant child; but then, embarrassed, I think better of it and grab his hand, and then Neil's.

In this fascinating chapter from his 1979 study of the Wrights, *Kill Devil Hill*, Harry Combs tells in detail the true nature of the brothers' accomplishment. He presents Orville and Wilbur not as humble bicycle mechanics who happened to stumble upon the secret of flight but as geniuses the equal of their contemporaries Edison and Bell. Although they stood upon the shoulders of such giants as Leonardo, Cayley, Lilienthal, and Chanute, their achievement was not strictly an evolutionary advance: It was as much a departure from the understanding of the past as it was a continuation.

HARRY COMBS with MARTIN CAIDIN

The Secret of the Wrights

For thousands of years men had envied the birds and sought to emulate their flight. Some of the best minds of history had worked on the problem without tangible results. The failures surely can be attributed to complex technical problems, but they cannot be blamed on a lack of materials necessary to construct a flying machine. Many people believed these were not available until the turn of the last century; this is far from the truth.

Far back in antiquity, there had been a sufficiency of perfectly suitable materials for the building of a glider. For several millennia, man could have built a flying machine without power — only a glider, to be sure, but still a flying machine. The Egyptians, the Greeks, and the Romans could have done it.

In the beginning of the eighteenth century, steam power had become available. A steam engine is not an *ideal* power plant for an airplane for many reasons, but nevertheless it *is* possible for someone to build a man-carrying airplane powered by a steam engine and fly it. In all the years of the steam engine's development and refinement, no one had done so.

By 1870, the internal-combustion engine had been developed; relatively efficient, controllable, lightweight power was now available. By 1890, there had been a century's worth of engineering achievements — railroads, steamships, mechanical reapers, electric lights, telephones, telegraphs, machine tools, calculators — to stimulate and support the inventor, and they were still coming thick and fast; today's technological explosion had barely started.

As we have noted, many of the men responsible for these marvels had turned their attention to the problems of mechanical flight. They had everything available to them that the Wright brothers possessed. Most the them had superior education, better training, better working facilities, and a vast advantage in financial resources — yet they all fell far short of the goal.

Why is it, then, that in this same arena, with the same forces to work with, the Wrights, in four and half years, succeeded brilliantly where others had failed?

The answer is not simple, but it is there — clear and yet profound.

In the first place, the Wrights identified, defined, and outlined completely the problem they were attacking, which no other inventor — contemporary or earlier — had ever done.

Many of their predecessors, from lack of adequate knowledge, labored under the impression that there were only *one* or *two* problems involved in mechanical flight, and that if they solved these, they would be successful. They hadn't perceived that there were many more snakes in the basket, any one of which could prevent them from succeeding.

The Wrights saw at once that each of the problems was a complex of many parts and approached each one in turn.

The Wrights, with amazing perception, immediately saw and defined the six problems whose solutions were necessary to the attainment of flight.

1. To emulate the birds, one must understand that there are two distinct types of bird flight, and that man for mechanical and gymnastic reasons could hope to copy only one. The darting flight of wing-flapping barn swallows, for instance, required impossible solutions to the problems of mechanical application and human reflexes. The second type, the soaring flight of the albatross or eagle, man could design and emulate, if he could understand its principles.

2. The solution of the secret of lift, with its problems of equilibrium, was related to, but different from, the third problem.

3. There had to be designed a satisfactory three-dimensional control system.

4. An apparatus strong enough to support the weight of a man, yet light enough to fly, must be constructed.

5. The power-to-weight ratio had to be correct.

6. The pilot had to achieve the necessary skill to fly the machine without being destroyed in the process.

But equally significant was the way the Wrights attacked each problem after identifying it. The first problem was to understand and identify the two main forms of bird flight.

The observations of Mouillard while he watched the flight of buzzards in the Egyptian desert confirmed what Wilbur and Orville had already agreed upon. In addition, they were further convinced by observations noted in *Animal Mechanism*, a famous work of E. J. Marey, which they had studied before writing to the Smithsonian.

Watching the barn swallows dart about a barn—whirring themselves in and out of the doors, perching on lofts and rafters, fluttering here and there with an accuracy and an astounding capability despite the potential turbulence of wind drafts, was confusing. It was obvious that the flapping of their wings sustained them. But how? And how did they maintain their extraordinary control?

Wilbur was able to see two vital things about the fluttering flight of a barn swallow. First, the construction of a mechanism that would reproduce the motions of these flapping wings with any degree of success would border on the impossible, and even if it could be constructed, it would weigh too much to raise itself aloft.

Second, and more difficult, was the understanding of the reflexes the swallows needed to maintain equilibrium during these extraordinary darting flights. Wilbur saw that the flight of a barn swallow was a series of gymnastic feats of the first magnitude, and that no man could hope to equal it or to maneuver a machine in this way.

There was another reason that, at this point, he had not completely understood, but that would develop later as the brothers expanded their experiments. This was simply that the required power-to-weight ratio for flapping wings was infinitely greater than would be required in rigid-wing flight. Leonardo da Vinci had made drawings of a proposed apparatus by which a strong man could flap wings and actuate the mechanism by both arms and legs, utilizing the full power of his body.

What Leonardo did not know was that it would take approximately 50 times the strength of a man to sustain himself in the air by the mechanism he had proposed. Few of us have any real conception of the relative strength of a bird and the power used to actuate his wings in comparison to his own weight. A man's pulse is approximately 70 beats per minute, that of a small sparrow will be perhaps 500 to 700. In other words, a sparrow's generation of energy, compared to a man's, is something on the order of 10 to 1.

To produce this enormous energy, the bird must consume approximately his own weight in food every day. If a man were to consume his own weight in food in a day, it would be his last day on earth; he has no way of converting energy at this rate.*

The soaring bird—the eagle, for example—was observed to keep its wings in a rigid position and soar in spirals in the sky, or hover along ridges, balancing adroitly with the winds and currents, but obviously exerting no flapping motion to sustain itself in the air. This was clearly the best approach for man to take, because it was apparent that it would be far better to construct a wing like that of a soaring bird, where no motion of its structure was required.

---

*It is interesting to note that many jet airplanes consume approximately their own weight in fuel every sixteen to eighteen hours—very much on the order of a bird's consumption.

Theories along this line had been advanced, notably by Lilienthal, though also by others, that involved tangential pressures developed by presenting a flat or even a curved surface toward moving air. That problem involved lift and equilibrium, and the Wrights attacked it later.

They had arrived at the correct conclusion: there was only one mode of bird flight they could hope to emulate. This is what Wilbur meant when he wrote to the Smithsonian that "I believe that simple flight at least is possible to man." Simple flight, of course, was the flight of rigid-winged birds, which sailed in some mysterious way on the currents of the air.

Now they came to the question of lift with equilibrium. It was apparent that if a square, thin surface were presented flat to the wind, it would offer maximum resistance, and that if it were turned edgeways to the wind, it would provide very little. Any angle between the vertical and the horizontal would result in some sort of tangential force — a force that would act in some direction other than that of the wind direction. They had reasoned from their kite-flying activities that the direction and strength of this tangential force could be calculated. A center of pressure — the point where the force acted — could be established, but it was quite obvious that the center of pressure would move. This they called the center-of-pressure travel, and assumed that it would move forward on the surface as the angle of the surface to the wind was decreased until the surface finally reached a zero or horizontal angle to the wind, in which case the center of pressure would finally arrive at the leading edge of the surface.

They assumed that the same conditions would apply to curved surfaces, and they were prepared to use the curved surfaces because Lilienthal and others had proven that curved surfaces had more lift than flat surfaces. No one really knew why, but birds' wings were curved and Lilienthal's glider wings were curved, and both provided lift.

This was logical, but would later lead the Wrights into dangerous traps. In any case, it was their conviction that a flat or curved surface could be so presented to the angle of the wind that, if the wind had sufficient velocity and if there was sufficient surface, enough lift could be generated to equal the weight of a man, and flight would be possible.

Also, if the center of pressure moved from the center point of the plane's surface area when the surface was at right angles to the wind, forward to the leading edge of the plane surface when that surface was slanted toward the wind, the problems of lift would have been solved, but the problem of equilibrium would be ever present. To balance the aircraft on that center of pressure would be like balancing on a tight rope in a strong wind. That plane surface would obviously wobble as the skis of a water skier wobble when presented at various angles to the water and at changing speeds. Of their center-of-pressure travel and the difficulty involved in balancing a glider or an airplane caught in its web, Wilbur had this to say: "The balancing of a gliding or flying machine is very simple in theory. It merely consists in causing the center of pressure to coincide with the center of gravity. But in actual practice there seems to be an almost boundless incompatibility of temper which prevents their remaining peaceably together for a single instant, so that the operator, who in this case acts as a peacemaker, often suffers injury to himself in attempting to bring them together." The Wrights saw the problem clearly for the first time.

They could also see that to structure from this flat or curved surface, which they would now call a wing, they had to find some way of balancing and counteracting the center-of-pressure travel. This balancing arrangement could not be in the form of a movable weight, for many reasons; rather, it had to be some sort of offsetting lifting surface. This is the function served by the horizontal part of the tail of a present-day airplane. The brothers came up with the idea of a small forward plane, to be built in an extended position ahead of their wing. At first they thought it could be secured in a fixed position, but they would later discover that it would have to be movable if it was to maintain the aircraft's balance. The first forward plane, which at the time, they chose to call the horizontal rudder but which we today would call a forward elevator (or canard), became the hallmark of their aircraft design. All Wright planes, in the early days, were distinguished by the forward plane, which extended ahead of the wings.

The fact that the Wrights had incorrectly analyzed the center-of-pressure travel, and its differences on flat surfaces versus curved surfaces, turned out, in the long run, to make no difference. This forward plane would be a happy accident of design, because, as we shall see, it would save their lives many times over.

Many people confuse the problem of lift and equilibrium with that of control. Here again the Wrights saw the difference. It was one thing to maintain balance as the center of pressure moved on the wing; it was quite another to control the motion of the aircraft in three dimensions.

The solution to this problem did not come easily. The brothers labored over it mightily and sought solutions to it no matter where they were or what they did. Their preoccupation was enough to cut them off at times from contact with the outside world. When they discussed the matter, it was in soul-searching or thundering arguments. They kept trying different methods, "talking" with their hands or scribbling out sketches, to see what the other thought about an idea or a suggestion. They were agreed in practice, at least, that if they could change the angle at which the wing attacked the air, and change it rapidly and with reliable response, they would gain an element of control. They devised a system of gears that would change the wing angle of a flying machine, but then, after weeks of work, had to conclude that it would be impossible for them to build a structure strong and rigid enough for the system and still light enough to fly. Their solutions seemed to follow a circle that always returned them to their point of starting — which meant no solution at all.

Then, in the summer of 1899, Wilbur waited impatiently for his brother to return to the house at 7 Hawthorn Street. The usually taciturn Wilbur fairly dragged Orville to a chair to force him to listen; no problem, of course, because each listened eagerly when the other had a new idea.

Orville listened as Wilbur, arms moving like windmill blades, his unusual excitement contagious, explained that he had solved the problems of lateral control. Not in a machine he had built, but in his mind, and with the manner of exchange of these two brothers, it was enough for Orville to grasp clearly what Wilbur proposed.

Years later, Orville would recall:

> A short time afterwards, one evening when I returned home with my sister and Miss Harriet Silliman, who was at that time a guest of my sister's in our home, Wilbur showed me a method of getting the same results as we had contemplated in our first idea without the structural defects of the original. He demonstrated the method by means of a small pasteboard box, which had two of the opposite ends removed. By holding the top forward corner and the rear lower corner of one end of the box between his thumb and forefinger and the rear upper corner and the lower forward corner of the other end of the box in like manner, and by pressing the corners together the upper and lower surface of the box were given a helicoidal twist, presenting the top and bottom surfaces of the box at different angles on the right and left sides. From this it was apparent that the wings of a machine of the [Octave] Chanute double-deck type, with the fore-and-aft trussing removed, could be warped in like manner, so that in flying the wings on the right and left sides could be warped so as to present their surfaces to the air at different angles of incidence and thus secure unequal lifts on the two sides.

For the next few weeks the bicycle shop saw nothing of Wilbur. He had been making sketches of this new design, trying to blend it with a biplane structure, which he and Orville had agreed was the strongest way to attain adequate surface with light weight. Now he had a plan: sets of wires run through small pulleys between the top and bottom wings would enable him to warp the wings in flight. This is essentially how the wings of today's aircraft are controlled to keep them level or to make them bank; the movable portion of the wing—the aileron—is still located at the tip of the wing and at the trailing edge but is hinged rather than bent.

If he was right, he explained to Orville, then they could achieve with an airplane wing what a bird did by flexing the outer parts, and especially the outer trailing edges, of its wings. He built a model of split bamboo, threads, and connecting trusses for bracing, shaping, and reshaping until he had worked out in practice what he had designed. But this wasn't enough. He studied every motion of his delicate model as he moved the warping system back and forth in his hands, jotting down notes, improving here and there, defining and refining. Then it was a feverish rush to build a flyable model that would enable him to test in flight what was, no matter how clever, nothing more than a theory on paper and a bamboo contraption in his workshop.

At this stage the brothers weren't even dreaming of flying a man. Wilbur was desperate to get into the air a model that incorporated wing warping* and could be flown as a kite to test wing warping as an effective flight control; he could at least then see what wing warping would accomplish. By unspoken agreement, since Wilbur had conceived the idea, Orville attended to the mundane and now frustrating daily chores of running the bicycle shop while Wilbur rushed ahead with the kite model. Six days after he had demonstrated his warping idea to Orville, on July 27, 1899, Wilbur was ready to test it. The model was impressive in size, with a wingspan of five feet. Its

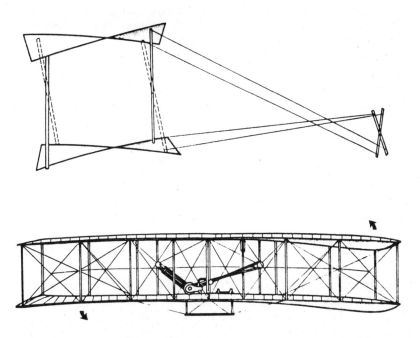

Wing warping: Top sketch is side-view diagram of the Wrights' biplane kite of 1899, indicating general arrangement of strings and effect of wing-warping movements. Operator maneuvered machine by actuating sticks (at right) with fore-and-aft wrist motion. Bottom drawing shows how wing-warping principle tested in the biplane kite worked when incorporated in the *Kitty Hawk Flyer* design. The drawing shows left-wing trailing edge (at right in sketch) warped down, thus producing more lift than the opposite wing and effecting a rolling motion to the right.

*The Wrights originally used the word *twisting*, or *helical twisting*, to define the changing of a wing's angle of incidence at the tip for lateral control. But the term *warping*, credited to Octave Chanute, was adopted by the Wrights and became generally accepted. Later the French translated it literally as "gauchir" and "gauchissement." In their French patent, however, the Wrights used *tordre* and *torsion* — meaning "to twist."

biplane wings measured a foot in width, or chord (chord is the wing measurement from leading to trailing edge), and the wings themselves were curved along their surface as Lilienthal's had been in his gliders. If this simple device was to work as Wilbur hoped, it would be an important step forward.

Much has been made of the predilection of the Wright brothers toward secrecy. It has been explained as stemming from their fear of having their ideas stolen by other experimenters, but in these early stages it was probably no more than a natural desire for privacy. Wilbur did not want to become a buffoon by flying the kite in the midst of gawking crowds. In their younger days, Orville had been a splendid designer, builder, and flier of kites, and had sent his contraptions soaring to great heights, but he was content to let Wilbur fly the experimental kite because Wilbur had originated the concept he would now test.

In a field to the west of the city Wilbur prepared to try out the system. The only witnesses were several schoolboys who had caught sight of him. This promised to be fun; when a man flew a kite, especially one this size, you could expect something really special.

The youngsters watched Wilbur reel out his control lines. They were puzzled by the new, different shape of the kite and by the way the well-dressed man worked on a series of lines. They saw the lines tugging and pulling at the wingtips of the kite, and the shapes bent to and fro, and they wondered at the way the double-winged kite went first this way, and then the other, almost as if the man with the strings were controlling its motion in the air.

Not too successfully, they may have decided, for something suddenly went wrong, and the boys looked up to see the big kite abruptly dip downward and rush wildly at them in a swift dive. No time to run—the boys threw themselves flat on the ground, hands clasped over heads as they readied themselves for the impact, but all they felt was a whistling sound as the kite rushed by above them, despite everything the man did to prevent that downward plunge.

Both Wilbur and his brother were pleased at the way the kite had responded to lateral control but were baffled by the way it had plunged. They had no explanation for this misbehavior, for while it was held aloft, taut at the end of its controlling string, it flew splendidly and answered every command of Wilbur's as he warped the wings. They had no way of knowing at the time that the problem lay not in their design, but in their continuing belief that Otto Lilienthal was the absolute authority on the best curvature for a wing. A wing may be curved in the form of an arc or in the form of a parabola, and there is danger inherent in both designs, but more so in one than the other. At that moment they did not appreciate that there was a killer hiding in the shape of the kite's wings, nor would they—until it threatened their own lives. At the moment, however, the overall results of the kite experiment were absolutely splendid. *They had proved that wing warping worked;* that it indeed provided lateral control by aerodynamic means and could be substituted for the pilot's shifting weight. There were air currents and forces yet to be reckoned with, but no one else had ever achieved satisfactory lateral control in either a model or a full-sized machine, and they had done so on that open field outside Dayton.

They were ready to make the next step, to move up to the ranks of Lilienthal, Pilcher, and all the others who had risked their lives in the air. Using their new design, with wing warping, they would now build a machine large enough, strong enough, and with enough lifting force, to bear the weight of a man.

Charlie's a great pilot, but he's lost his edge, the young man's edge which consists simply of his belief that he cannot die — not just yet, anyway. This harrowing tale was published in 1946, in *Over to You: Ten Stories of Fliers and Flying*.

ROALD DAHL

Death of an Old Man

Oh God, how I am frightened.

Now that I am alone I don't have to hide it; I don't have to hide anything any longer. I can let my face go because no one can see me; because there's 21,000 feet between me and them and because now that it's happening again I couldn't pretend any more even if I wanted to. Now I don't have to press my teeth together and tighten the muscles of my jaw as I did during lunch when the corporal brought in the message; when he handed it to Tinker and Tinker looked up at me and said, "Charlie, it's your turn. You're next up." As if I didn't know that. As if I didn't know that I was next up. As if I didn't know it last night when I went to bed, and at midnight when I was still awake and all the way through the night, at one in the morning and at two and three and four and five and six and seven o'clock when I got up. As if I didn't know it while I was dressing and while I was having breakfast and while I was reading the magazines in the mess, playing shove-halfpenny in the mess, reading the notices in the mess, playing billiards in the mess. I knew it then and I knew it when we went in to lunch, while we were eating that mutton for lunch. And when the corporal came into the room with the message — it wasn't anything at all. It wasn't anything more than when it begins to rain because there is a black cloud in the sky. When he handed the paper to Tinker I knew what Tinker was going to say before he had opened his mouth. I knew exactly what he was going to say.

So that wasn't anything either.

But when he folded the message up and put it in his pocket and said, "Finish your pudding. You've got plenty of time," that was when it got worse, because I knew for certain then that it was going to happen again, that within half an hour I would be strapping myself in and testing the engine and signaling to the airmen to pull away the chocks. The others were all sitting around eating their pudding; mine was still on my plate in front of me, and I couldn't take another mouthful. But it was fine when I tightened my jaw muscles and said, "Thank God for that. I'm tired of sitting around here picking my nose." It was certainly fine when I said that. It must have sounded like any of the others just before they started off. And when I got up to leave the table and said, "See you at tea time," that must have sounded all right too.

But now I don't have to do any of that. Thank Christ I don't have to do that now. I can just loosen up and let myself go. I can do or say anything I want so long as I fly this airplane properly. It didn't use to be like this. Four years ago it was wonderful. I loved doing it because it was exciting, because the waiting on the aerodrome was nothing more than the waiting before a football game or before going in to bat; and three years ago it was all right too. But then always the three months of resting and the going back again and the resting and going back; always going back and always getting away with it, everyone saying what a fine pilot, no one knowing what a near thing it was that time near Brussels and how lucky it was that time over Dieppe and how bad it was that other time over Dieppe and how lucky and bad and scared I've been every minute of every trip every week this year. No one knows that. They all say, "Charlie's a great pilot," "Charlie's a born flier," "Charlie's terrific."

I think he was once, but not any longer.

Each time now it gets worse. At first it begins to grow upon you slowly, coming upon you slowly, creeping up on you from behind, making no noise, so that you do not turn round and see it coming. If you saw it coming, perhaps you could stop it, but there is no warning. It creeps closer and closer, like a cat creeps closer stalking a sparrow, and then when it is right behind you, it doesn't spring like the cat would spring; it just leans forward and whispers in your ear. It touches you gently upon the shoulder and whispers to you that you are young, that you have a million things to do and a million things to say, that if you are not careful you will buy it, that you are almost certain to buy it sooner or later, and that when you do you will not be anything any longer; you will just be a charred corpse. It whispers to you about how your corpse will look when it is charred, how black it will be and how it will be twisted and brittle, with the face black and the fingers black and the shoes off the feet because the shoes always come off the feet when you die like that. At first it whispers to you only at night, when you are lying awake in bed at night. Then it whispers to you at odd moments during the day, when you are doing your teeth or drinking a beer or when you are walking down the passage; and in the end it becomes so that you hear it all day and all night all the time.

There's Ijmuiden. Just the same as ever, with the little knob sticking out just beside it. There are the Frisians, Texel, Vlieland, Terschelling, Ameland, Juist and Norderney. I know them all. They look like bacteria under a microscope. There's the Zuider Zee, there's Holland, there's the North Sea, there's Belguim, and there's the world; there's the whole bloody world right there, with all the people who aren't going to get killed and all the houses and the towns and the sea with all the fish. The fish aren't going to get killed either. I'm the only one that's going to get killed. I don't want to die. Oh God, I don't want to die. I don't want to die today anyway. And it isn't the pain. Really it isn't the pain. I don't mind having my leg mashed or my arm burnt off; I swear to you that I don't mind that. But I don't want to die. Four years ago I didn't mind. I remember distinctly not minding about it four years ago. I didn't mind about it three years ago either. It was all fine and exciting; it always is when it looks as though you may be going to lose, as it did then. It is always fine to fight when you are going to lose everything anyway, and that was how it was four years ago. But now we're going to win. It is so different when you are going to win. If I die now I lose fifty years of life, and I don't want to lose that. I'll lose anything except that because that would be all the things I want to do and all the things I want to see; all the things like going on sleeping with Joey. Like going home sometimes. Like walking through a wood. Like pouring out a drink from a bottle. Like looking forward to weekends and like being alive every day every year for fifty years. If I die now I will miss all that, and I will miss everything else. I will miss the things that I don't know about. I think those are really the things I am frightened of missing. I think the reason I do not want to die is because of the things I hope will happen. Yes, that's right. I'm sure that's right. Point a revolver at a tramp, at a wet shivering tramp on the side of the road and say, "I'm going to shoot you," and he will cry, "Don't shoot. Please don't shoot." The tramp clings to life because of the things he hopes will happen. I am clinging to it for the same reason; but I have clung for so long now that I cannot hold on much longer. Soon I will have to let go. It is like hanging over the edge of a cliff, that's what it is like; and I've been hanging on too long now, holding on to the top of the cliff with my fingers, not being able to pull myself back up, with my fingers getting more and more tired, beginning to hurt and to ache, so that I know that sooner or later I will have to let go. I dare not cry out for help; that is one thing that I dare not do; so I go on hanging over the side of this cliff, and as I hang I keep kicking a little with my feet against the side of the cliff, trying desperately to find a foothold, but it is steep and smooth like the side of a ship, and there isn't any foothold. I am kicking now, that's what I am doing. I am kicking against the smooth side of the cliff, and there isn't any foothold. Soon I shall have to let go. The longer I hang on the more certain I am of that, and so each hour, each day, each night, each week, I become more and more frightened. Four years ago I wasn't hanging over the edge like this. I was running about in the field above, and although I knew that there was a cliff somewhere and that I might fall over it, I did not mind. Three years ago it was the same, but now it is different.

I know that I am not a coward. I am certain of that. I will always keep going. Here I am today, at two o'clock in the afternoon, sitting here flying a course of 135 at 360 miles an hour and flying well; and although I am so frightened that I can hardly think, yet I am going on to do this thing. There was never any question of not going or of turning back. I would rather die than turn back. Turning back never enters into it. It would be easier if it did. I would prefer to have to fight that than to have to fight this fear.

There's Wassalt. Little camouflaged group of buildings and great big camouflaged aerodrome, probably full of one-o-nines and one-nineties. Holland looks wonderful. It must be a lovely place in the summer. I expect they are haymaking down there now. I expect the German soldiers are watching the Dutch girls haymaking. Bastards. Watching them haymaking, then making them come home with them afterwards. I would like to be haymaking now. I would like to be haymaking and drinking cider.

The pilot was sitting upright in the cockpit. His face was nearly hidden by his goggles and by his oxygen mask. His right hand was resting lightly upon the stick, and his left hand was forward on the throttle. All the time he was looking around him into the sky. From force of habit his head never ceased to move from one side to the other, slowly, mechanically, like clockwork, so that each moment almost, he searched every part of the blue sky, above, below and all around. But it was into the light of the sun itself that he looked twice as long as he looked anywhere else; for that is the place where the enemy hides and waits before he jumps upon you. There are only two places in which you can hide yourself when you are up in the sky. One is in a cloud and the other is in the light of the sun.

He flew on; and although his mind was working upon many things and although his brain was the brain of a frightened man, yet his instinct was the instinct of a pilot who is in the sky of the enemy. With a quick glance, without stopping the movement of his head, he looked down and checked his instruments. The glance took no more than a second, and like a camera can record a dozen things at once with the opening of a shutter, so he at a glance recorded with his eyes his oil pressure, his petrol, his oxygen, his rev counter, his boost and his airspeed, and in the same instant almost, he was looking up again into the sky. He looked at the sun, and as he looked, as he screwed up his eyes and searched into the dazzling brightness of the sun, he thought that he saw something. Yes, there it was; a small black speck moving slowly across the bright surface of the sun, and to him the speck was not a speck but a life-size German pilot sitting in a Focke Wulf which had cannon in its wings.

He knew that he had been seen. He was certain that the one above was watching him, taking his time, sure of being hidden in the brightness of the sun, watching the Spitfire and waiting to pounce. The man in the Spitfire did not take his eyes away from the small speck of black. His head was quite still now. He was watching the enemy, and as he watched, his left hand came away from the throttle and began to move delicately around the cockpit. It moved quickly and surely, touching this thing and that, switching on his reflector sight, turning his trigger button from "safe" over to "fire" and pressing gently with his thumb upon a lever which increased, ever so slightly, the pitch of the airscrew.

There was no thought in his head now save for the thought of battle. He was no longer frightened or thinking of being frightened. All that was a dream, and as a sleeper who opens his eyes in the morning and forgets his dream, so this man had seen the enemy and had forgotten that he was frightened. It was always the same. It had happened a hundred times before, and now it was happening again. Suddenly, in an instant, he had become cool and precise, and as he prepared himself, as he made ready his cockpit, he watched the German, waiting to see what he would do.

This man was a great pilot. He was great because when the time came, whenever the moment arrived, his coolness was great and his courage was great, and more than anything else his instinct was great, greater by far than his coolness or his courage or his experience. Now he eased open the throttle and pulled the stick gently backward, trying to gain height, trying to gain a little of the five-thousand-feet advantage which the German had over him. But there was not much time. The Focke Wulf came out of the sun with its nose down and it came fast. The pilot saw it coming and he kept going straight on, pretending that he had not seen it, and all the time he was looking over his shoulder, watching the German, waiting for the moment to turn. If he turned too soon, the German would turn with him, and he would be duck soup. If he turned too late, the German would get him anyway provided that he could shoot straight, and he would be duck soup then too. So he watched and waited, turning his head and looking over his shoulder, judging his distance; and as the German came within range, as he was about to press his thumb upon the trigger button, the pilot swerved. He yanked the stick hard back and over to the left, he kicked hard with his left foot upon the rudder-bar, and like a leaf which is caught up and carried away by a gust of wind, the Spitfire flipped over on to its side and changed direction. The pilot blacked out.

As his sight came back, as the blood drained away from his head and from his eyes, he looked up and saw the German fighter way ahead, turning with him, banking hard, trying to turn tighter and tighter in order to get back on the tail of the Spitfire. The fight was on. "Here we go," he said to himself. "Here we go again," and he smiled once, quickly, bacause he was confident and because he had done this so many times before and because each time he had won.

The man was a beautiful pilot. But the German was good too, and when the Spitfire applied a little flap in order to turn in tighter circles, the Focke Wulf appeared to do the same, and they turned together. When the Spitfire throttled back suddenly and got on his tail, the Focke Wulf half-rolled and dived out and under and was away, pulling up again in a loop and rolling off the top, so that he came in again from behind. The Spitfire half-rolled and dived away, but the Focke Wulf anticipated him, and half-rolled and dived with him, behind him on his tail, and here he took a quick shot at the

Spitfire, but he missed. For at least fifteen minutes the two small aircraft rolled and dived around each other in the sky. Sometimes they would separate, wheeling around and around in tight turns watching one another, circling and watching like two boxers circling each other in the ring, waiting for an opening or for the dropping of a guard; then there would be a stall-turn and one would attack the other, and the diving and the rolling and the zooming would start all over again.

All the time the pilot of the Spitfire sat upright in his cockpit, and he flew his aircraft not with his hands, but with the tips of his fingers, and the Spitfire was not a Spitfire but a part of his own body; the muscles of his arms and legs were in the wings and in the tail of the machine so that when he banked and turned and dived and climbed he was not moving his hands and his legs, but only the wings and the tail and the body of the airplane; for the body of the Spitfire was the body of the pilot, and there was no difference between the one and the other.

So it went on, and all the while, as they fought and as they flew, they lost height, coming down nearer and nearer to the fields of Holland, so that soon they were fighting only 3,000 feet above the ground, and one could see the hedges and the small trees and shadows which the small trees made upon the grass.

Once the German tried a long shot, from a thousand yards, and the pilot of the Spitfire saw the tracer streaming past in front of the nose of his machine. Once, when they flew close past each other, he saw, for a moment, the head and shoulders of the German under the glass roof of his cockpit, the head turned toward him, with the brown helmet, the goggles, the nose and the white scarf. Once when he blacked out from a quick pullout, the blackout lasted longer than usual. It lasted maybe five seconds, and when his sight came back, he looked quickly around for the Focke Wulf and saw it half a mile away, flying straight at him on the beam, a thin inch-long black line which grew quickly, so that almost at once it was no longer an inch, but an inch and a half, then two inches, then six and then a foot. There was hardly any time. There was a second or perhaps two at the most, but it was enough because he did not have to think or to wonder what to do; he had only to allow his instinct to control his arms and his legs and the wings and the body of the airplane. There was only one thing to do, and the Spitfire did it. It banked steeply and turned at right angles toward the Focke Wulf, facing it and flying straight toward it for a head-on attack.

The two machines flew fast toward each other. The pilot of the Spitfire sat upright in his cockpit, and now, more than ever, the aircraft was a part of his body. His eye was upon the reflector sight, the small yellow electric-light dot which was projected up in front of the windshield, and it was upon the thinness of the Focke Wulf beyond. Quickly, precisely, he moved his aircraft a little this way and that, and the yellow dot, which moved with the aircraft, danced and jerked this way and that, and then suddenly it was upon the thin line of the Focke Wulf and there it stayed. His right thumb in the leather glove felt for the firing button; he squeezed it gently, as a rifleman squeezes a trigger, his guns fired, and at the same time, he saw the small spurts of flame from the cannon in the nose of the Focke Wulf. The whole thing, from beginning to end, took perhaps as long as it would take you to light a cigarette. The German pilot came straight on at him and he had a sudden, vivid, colorless view of the round nose and the thin outstretched wings of the Focke Wulf. Then there was a crack as their wing tips met, and there was a splintering as the port wing of the Spitfire came away from the body of the machine.

The Spitfire was dead. It fell like a dead bird falls, fluttering a little as it died; continuing in the direction of its flight as it fell. The hands of the pilot, almost in a single movement, undid his straps, tore off his helmet and slid back the hood of the cockpit; then they grasped the edges of the cockpit and he was out and away, falling, reaching for the ripcord, grasping it with his right hand, pulling on it so that his parachute billowed out and opened and the straps jerked him hard between the fork of his legs.

All of a sudden the silence was great. The wind was blowing on his face and in his hair and he reached up a hand and brushed the hair away from his eyes. He was about a thousand feet up, and he looked down and saw flat green country with fields and hedges and no trees. He could see some cows in the field below him. Then he looked up, and as he looked, he said, "Good God," and his right hand moved quickly to his right hip, feeling for his revolver which he had not brought with him. For there, not more than five hundred yards away, parachuting down at the same time and at the same height, was another man, and he knew when he saw him that it could be only the German pilot. Obviously his plane had been damaged at the same time as the Spitfire in the collision. He must have got out quickly too; and now here they were, both of them parachuting down so close to each other that they might even land in the same field.

He looked again at the German, hanging there in his straps with his legs apart, his hands above his head grasping the cords of the parachute. He seemed to be a small man, thickly built and by no means young. The German was looking at him too. He kept looking, and when his body swung around the other way, he turned his head, looking over his shoulder.

So they went on down. Both men were watching each other, thinking about what would happen soon, and the German was the king because he was landing in his own territory. The pilot of the Spitfire was coming down in enemy country; he would be taken prisoner, or he would be killed, or he would kill the German, and if he did that, he would escape. I will escape anyway, he thought. I'm sure I can run faster than the German. He does not look as though he could run very fast. I will race him across the fields and get away.

The ground was close now. There were not many seconds to go. He saw that the German would almost certainly land in the same field as he, the field with the cows. He looked down to see what the field was like and whether the hedges were thick and whether there was a gate in the hedge, and as he looked, he saw below him in the field a small pond, and there was a small stream running through the pond. It was a cow-drinking pond, muddy round the edges and muddy in the water. The pond was right below him. He was no more than the height of a house above it and he was dropping fast; he was dropping right into the middle of the pond. Quickly he grasped the cords above his head and tried to spill the parachute to one side so that he would change direction, but he was too late; it wasn't any good. All at once something brushed the surface of his brain and the top of his stomach, and the fear which he had forgotten in the fighting was upon him again. He saw the pond and the black surface of the water of the pond, and the pond was not a pond, and the water was not water; it was a small black hole in the surface of the earth which went on down and down for miles and miles, with steep smooth sides like the sides of a ship, and it was so deep that when you fell into it, you went on falling and falling and you fell forever. He saw the mouth of the hole and the deepness of it, and he was only a small brown pebble which someone had picked up and thrown into the air so that it would fall into the hole. He was a pebble which someone had picked up in the grass of the field. That was all he was and now he was falling and the hole was below him.

Splash. He hit the water. He went through the water and his feet hit the bottom of the pond. They sank into the mud on the bottom and his head went under the water, but it came up again and he was standing with the water up to his shoulders. The parachute was on top of him; his head was tangled in a mass of cords and white silk and he pulled at them with his hands, first this way and then that, but it only got worse, and the fear got worse because the white silk was covering his head so that he could see nothing but a mass of white cloth and a tangle of cords. Then he tried to move toward the bank, but his feet were stuck in the mud; he had sunk up to his knees in the mud. So he fought the parachute and the tangled cords of the parachute, pulling at them with his hands and trying to get them clear of his head; and as he did so he heard the sound of footsteps running on the grass. He heard the noise of the footsteps coming closer and the German must have jumped, because there was a splash and he was knocked over the by weight of a man's body.

He was under the water, and instinctively he began to struggle. But his feet were still stuck in the mud, the man was on top of him and there were hands around his neck holding him under and squeezing his neck with strong fingers. He opened his eyes and saw brown water. He noticed the bubbles in the water, small bright bubbles rising slowly upward in the brown water. There was no noise or shouting or anything else, but only the bright bubbles moving upward in the water, and suddenly, as he watched them, his mind became clear and calm like a sunny day. I won't struggle, he thought. There is no point in struggling, for when there is a black cloud in the sky, it is bound to rain.

He relaxed his body and all the muscles in his body because he had no further wish to struggle. How nice it is not to struggle, he thought. There is no point in struggling. I was a fool to have struggled so much and for so long; I was a fool to have prayed for the sun when there was a black cloud in the sky. I should have prayed for rain; I should have shouted for rain. I should have shouted, Let it rain, let it rain in solid sheets and I will not care. Then it would have been easy. It would have been so easy then. I have struggled for five years and now I don't have to do it any more. This is so much better; this is ever so much better, because there is a wood somewhere that I wish to walk through, and you cannot walk struggling through a wood. There is a girl somewhere that I wish to sleep with, and you cannot sleep struggling with a girl. You cannot do anything struggling; especially you cannot live struggling, and so now I am going to do all the things that I want to do, and there will be no more struggling.

See how calm and lovely it is like this. See how sunny it is and what a beautiful field this is, with the cows and the little pond and the green hedges with primroses growing in the hedges. Nothing will worry me any more now, nothing nothing nothing; not even that man splashing in the water of the pond over there. He seems very puffed and out of breath. He seems to be dragging something out of the pond, something heavy. Now he's got it to the side, and he's pulling it up on to the grass. How funny; it's a body. It's a body of a man. As a matter of fact, I think it's me. Yes, it is me. I know it is, because of that smudge of yellow paint on the front of my flying suit. Now he's kneeling down, searching in my pockets, taking out my money and my identification card. He's found my pipe and the letter I got this morning from my mother. He's taking off my watch. Now he's getting up. He's going away. He's going to leave my body behind, lying on the grass beside the pond. He's walking quickly away across the field toward the gate. How wet and excited he looks. He ought to relax a bit. He ought to relax like me. He can't be enjoying himself that way. I think I will tell him.

"Why don't you relax a bit?"

Goodness, how he jumped when I spoke to him. And his face; just look at his face. I've never seen a man look as frightened as that. He's starting to run. He keeps looking back over his shoulder, but he keeps on running. But just look at his face; just look how unhappy and frightened he is. I do not want to go with him. I think I'll leave him. I think I'll stay here for a bit. I think I'll go along the hedges and find some primroses, and if I am lucky I may find some white violets. Then I will go to sleep. I will go to sleep in the sun.

The fate of democracy hung in the balance during that momentous summer of 1940 when the Luftwaffe brought the war to British skies and rained bombs on British soil. In this brief section of *Fighter: The True Story of the Battle of Britain*, Len Deighton—who has described in several bestselling novels what it meant to be an airman in those days—introduces us to three RAF heroes.

LEN DEIGHTON

Three Flight Commanders

JULY 9

Al Deere, from Wanganui, New Zealand, had traveled halfway across the world in order to join the RAF in 1937. By July 1940 he was a flight commander with 54 Squadron, and after a month of intensive air fighting he had got his DFC from the hands of the King, at a ceremony held at Hornchurch airfield. By now he was as experienced as any fighter pilot that the British had.

On July 9 he was leading a formation on his fourth flight of the day when they found a German rescue floatplane, painted white with eight large red crosses, flying at wave height. It was escorted by a dozen Bf 109's, flying close behind it.

While one section attacked the floatplane, Deere dived upon the Messerschmitts, which split into two formations, climbing steeply to right and left respectively, turning as they went. Deere remembered this tactic and later used it with some success. Now Deere's flyers broke and individual combats started. Deere noted with satisfaction the way the new De Wilde bullets made "small dancing yellow flames" as they exploded against the enemy fighter. He found this a valuable way of judging the effect of his gunfire.

> I soon found another target. About 3,000 yards directly ahead of me, and at
> the same level, a Hun was just completing a turn preparatory to reentering
> the fray. He saw me almost immediately and rolled out of his turn toward

me so that a head-on attack became inevitable. Using both hands on the control column to steady the aircraft and thus keep my aim steady, I peered through the reflector sight at the rapidly closing enemy aircraft. We opened fire together, and immediately a hail of lead thudded into my Spitfire. One moment the Messerschmitt was a clearly defined shape, its wingspan nicely enclosed within the circle of my reflector sight, and the next it was on top of me, a terrifying blur which blotted out the sky ahead. Then we hit.

The crash snatched the control column out of Deere's hands and the cockpit harness bit painfully deep into his shoulders. The engine was vibrating, and the control column was jumping backwards and forwards. As Deere watched, the engine gave forth smoke and flames, and before he could switch off the ignition the propeller stopped. Now Deere could see that its blades were bent double: the Bf 109 had scraped along the top of his Spitfire.

Unable to get his hood open, Deere coaxed the aircraft into a glide toward the distant coastline, while he struggled to get out. With skill to match his amazing good fortune he brought the wrecked aircraft down into a field, very near to Manston airfield. Still unable to open his hood, he smashed his way out of it, his "bare hands wielding the strength of desperation," he said. He got clear of the wreckage, which burned brightly as the bullets exploded. "Won't you come in and have a cup of tea?" said a woman coming out of a nearby farmhouse.

"Thank you, I will," said Deere, "but I would prefer something stronger if you've got it."

Modestly Deere got back to his squadron expecting nothing more than a couple of days off. But they were so short of pilots that his commander asked him to fly again immediately. The squadron had lost two pilots to the Messerschmitts and when Deere got to dispersal there were only four Spitfires serviceable. "You needn't expect to fly this morning," he was told.

"I'm in no hurry," said Deere.

JULY 10

The Luftwaffe needed regular weather reports and more photo coverage of its targets. Not realizing what a sitting duck lone aircraft were for radar, the Germans came one at a time, flying out over the North Sea to what they hoped was an undefended landfall. Often it proved fatal for the German crew, but sometimes there were surprises. On July 10, a Do 17 looking for a convoy off the North Foreland had no less than an entire *Gruppe* of JG 51 flying escort on it. The other actions of the day were no more than skirmishes but by the end of it Fighter Command had flown over 600 sorties.*

*A sortie is a unit by which air activity is measured. A sortie equals one aircraft making one operational flight. Therefore ten sorties can mean ten aircraft flew one mission or one aircraft flew ten missions.

July 11

On this day there was an endless stream of lone German aircraft. The RAF responded by sending lone aircraft to meet them. Often the squadron commanders reserved that job for themselves. Not long after dawn, for instance, Peter Townsend—the commander of 85 Squadron at Martlesham—was at the controls of his Hurricane VY-K, climbing out of ground mist into low gray cloud and heavy rain. The controller's voice took him up to 8,000 feet, where, in cloud, he made a perfect interception.

This Dornier Do 17M was Y5+GM from *Kanalkampfführer* Fink's own *Geschwader*, II/KG 2, the *Holzhammer*. It came to England in a wide sweep out over the North Sea, reaching the coast near Lowestoft. Forbidden to bomb the mainland in daytime, after what one of the crew described as a little "sightseeing" they dropped their ten tiny (50-kg) bombs over some shipping in the harbor.

The crew had mixed feelings about the cloud and rain; it made the pilot's job more difficult, and reduced visibility for the bomb-aimer and gunner, but it was reassuring to think that any RAF fighters would be having the same sort of problems in locating them. The Germans were content, and as the nose of the Dornier turned for home the crew began singing "Good-bye, Johnny. . . ." The melody was interrupted by the sudden shout of a gunner, Werner Borner, "*Achtung, Jäger!*"

The *Jäger* was Peter Townsend, who could hardly see through his rainswept windscreen, and so slid open the cockpit cover to put his head out into the rainstorm. He had not had to rely on visibility until the last few moments, for the single Dornier had provided the radar plotters with a blip on the cathode-ray tube no more difficult to interpret than the ones set up as prewar exercises.

And Townsend was a peacetime pilot, a flyer of great skill and experience. His eight Browning machine guns raked the bomber. Inside the bomber there were "bits and pieces everywhere: blood-covered faces, the smell of cordite, all the windows shot up." Of the crew, the starboard rear gunner was hit in the head and fell to the floor. A second later another member of the crew—hit in head and throat—fell on top of him. There was blood everywhere. But "our good old Gustav Marie was still flying," remembered one of the crew. Townsend had put 220 bullets into the Dornier but it got home to Arras, and all the crew lived to count the bullet holes.

The German bombers were robust enough to endure terrible amounts of gunfire, especially small-caliber gunfire. Their strength did not depend upon bracing wires and wooden spars: these metal bombers had armor protection, with some of the vital mechanical parts duplicated. Even more valuable were the self-sealing fuel tanks. Of a very simple layered construction, they had a crude-rubber middle layer. As a puncture allowed fuel to spill, the crude rubber dissolved, swelled, and sealed the hole. The events of the day were to prove how effective these devices were in getting damaged bombers home.

Not only did Townsend's machine-gun bullets fail to shoot down his Dornier but a lucky shot from one of its machine guns hit his Hurricane's coolant system. The engine stopped when still twenty miles from the English coast. Townsend took to his parachute and was fished out of the ocean by a trawler that sailed into a minefield to reach him.

A little later, another squadron commander found another Dornier. This time the commander was the remarkable Douglas Bader. This peacetime fighter pilot had had both legs amputated after a flying accident but when war began he had been allowed to rejoin the RAF and fly once more as a fighter pilot. Already he had caught up with his contemporaries, and at the end of June he'd been made a Squadron Leader and assigned to command 242 (Canadian) Fighter Squadron. This squadron was flying old Mark I Hurricanes with two-blade fixed-pitch propellers. It consisted of Canadians who were serving in the RAF (it was not an RCAF unit) and had seen some air fighting in France. The squadron was deficient in equipment and morale when Bader arrived. He was now in the process of remedying those defects.

On the morning of July 11 at about seven o'clock Bader answered the phone in the dispersal hut near the aircraft. There was a single "bandit" flying up the coastline near Cromer and the Controller wanted a flight of Hurricanes to intercept it. Bader looked at the low cloud and decided that the Hurricanes would not be able to form up, so he would go alone. It was a significant decision from the man who later became the most enthusiastic proponent of "big wings," in which the fighters went to battle in large formations. It is interesting, too, to record that Bader (who was later to urge that the controllers should advise rather than order the fighter pilots about the enemy) this day found his victim without the assistance of the radar plotters.

His victim was a Dornier Do 17 of *Wetterkundungsstaffel* (weather reconnaissance unit) 261. It had already fought off two Spitfires. One of them—flown by the commander of 66 Squadron—had been damaged in the oil tank.

When Bader found his Dornier it was just under the cloud base at about 1,000 feet. Methodically Bader closed without being spotted until he was about 250 yards behind; then the German rear gunner opened fire. Bader fired two bursts as the Dornier turned to face back the way it had come and made a shallow climb until disappearing into the cloud. Cursing, Bader flew back to Coltishall and reported that his Dornier had escaped, but a few minutes later the telephone told him that the plane had crashed into the sea just a short time after his action. Modestly Bader described this as "a lucky start for the new CO of 242 Squadron," but there was no doubt that his success had come from skill and experience. There could be no such successes stemming from luck alone.

What a splendid career. In 1922, Jimmy Doolittle became the first person to fly across the United States in under twenty-four hours. Later he became one of the great stunt fliers, recording the first outside loop. He was a record-breaking racer who took the Schneider Cup in 1925, the Bendix Trophy in 1931, and the Thompson Trophy in 1932. He was a pioneer in instrument flying, taking off and landing blind in September 1929. And, of course, he led a flight of B-25 bombers in the Tokyo raid of April 18, 1942, which did so much to restore confidence to a badly battered American public. In this 1974 interview with Robert S. Gallagher, Lieutenant General Doolittle speaks spellbindingly of his years as a pioneer army aviator.

JIMMY DOOLITTLE

"I Am Not a Very Timid Type . . ."

I suppose the most obvious question to ask a pilot is how did you become interested in aviation?

In the winter of 1909–1910 I saw the first air show that took place on the West Coast at old Dominguez Flying Field, near my home in Los Angeles. I was very impressed with the airplanes of that day, even though they were quite frail and of very little performance. Well, I was at that time an avid reader of *Popular Mechanics*, which about two years later published an article about how to make a glider with sticks and wire and unbleached muslin. So I made a small biplane glider and took it to a small nearby cliff and jumped. Unfortunately the cruciform tail hit the edge of the cliff, and the glider came down rather abruptly. I wasn't badly hurt. but the glider was pretty badly broken up. I rebuilt it. Then I tied it to the rear of a friend's automobile. Using my legs as the landing gear, I ran behind the car as fast as I could and leaped into the air.

What happened?

This time the glider came down on its nose and was completely washed out. But about this time *Popular Mechanics* came out with a picture of Alberto Santos-Dumont's monoplane, the *Demoiselle*. So I gathered up the pieces of my glider and built a monoplane. I saved up my money and bought a secondhand motorcycle engine to power it. Probably the luckiest thing that happened was that a storm came along and blew my airplane a block away and completely wrecked it. It would have been a very dangerous craft to fly, I am quite sure.

Did that put an end to your early career as an aviator?

Until I got into World War I, in 1917, yes. You know, there is an organization called the Early Birds, which is made up of people who before that war built their own planes and learned to fly them. Recently they invited me to join, but I had to decline their invitation, because by no stretch of the human imagination can I claim that I ever achieved controlled flight in my boyhood experiments. . . .

Actually you started out to become an engineer, didn't you?

From the time I was a very young fellow, I knew I wanted to do two things. I wanted to build things, and I wanted to see the world. It seemed to me that the best way to build things was to be an engineer, and the best way to see the world was to be the kind of engineer that went to different parts of the world. In those days that meant either a civil or a mining engineer. I decided to become a mining engineer. . . .

[Nineteen seventeen] was the year you first got into aviation.

I was working in the mines that summer, at the Comstock Lode in Virginia City, Nevada. Oddly enough, although it was only a couple of hundred miles, I guess, from San Francisco, we were practically isolated from the outside world. It wasn't until I returned to college that I first learned we were really at war. And immediately I came to the realization that I wanted to participate. I looked around at the various military services one could get into, and the thought reawakened my desire to fly. You see, I have always been, to some degree, a loner. I enjoy being with a few intimate friends, but I have always been able to enjoy life even if I am alone. So I naturally went into fighter-pilot aviation, because there is a basic difference between the fighter pilot and the bomber pilot. The fighter pilot is almost always a rugged individualist, whereas the bomber pilot is more inclined to be a team player. For much the same reasons you will notice that a fighter pilot usually does a superb job in wartime and does not adapt to peacetime activities as well as the bomber pilot.

Well, at the beginning of World War I did you have much of an option between fighters and bombers?

You had an option in that you were permitted to express a preference. Only the pilots who seemed to be the most apt were normally taken into fighter training. There was a great deal more latitude then, of course, because there weren't as many bombers. At that time the Army Air Service was considered a defensive arm, to be used mainly for observation and reconnaissance purposes. Both the army and the navy were greatly opposed to building long-range bombers, and this made it very difficult to procure the necessary funds for research and development. We did have the twin-engine Martin bombers, and then around 1923 or 1924 the Barling bomber was built. It was a very large airplane, flown at Wright Field by Harold Harris, but it was so heavy and so underpowered that its poor performance made the concept of long-range heavy bombers look bad. If my memory is correct, in the late twenties Boeing built a much larger bomber which they called the B-15, but it had certain problems, and the army wouldn't buy it, so Boeing was permitted to sell it, as I recall, to the Japanese.

What did the Japanese do with it?

I would imagine that it was very useful to them for two reasons. First, to get some experience on how to build a big airplane, and second, to find out what made it unsatisfactory for American use. You learn as much from your failures, if you study them, as you learn from your successes.

Since you were trained as a fighter pilot, do you find it paradoxical that your best-known exploit in World War II was at the controls of a big bomber?

No. During the war I flew both bombers and fighters in North Africa and Italy. I did it for two reasons. First, I wanted to know as much and hopefully more about the equipment than the men who were flying it. Second, when I visited one of the bases in my command, I would arrive in one of their aircraft as a courtesy to my men. . . .

Where did you receive your flight training?

At Rockwell Field in San Diego. Jo [Josephine Daniels] and I were married the day before Christmas in 1917, and she went with me to San Diego. The training was much different in those days. We were in a terrible hurry to get pilots trained, and, of course, the airplanes were very simple. We flew the old Curtiss JN-4 or Jenny, and we practiced things like grass cutting, which meant holding the plane about three feet off the ground. After seven hours of flying with my instructor, good old Mr. [Charles] Todd, I soloed. It lasted thirty-nine minutes.

Would you tell me about the Jenny?

It was a very fine training plane, a two-seater, in which you could do almost anything that your skill permitted you to do. It was also quite a safe airplane to fly, although, with the OX-5 engine we had at that time, somewhat underpowered. I remember on one occasion landing in the Imperial Valley with an OX-5 Jenny. When I tried to take off, I found that I could not get the plane airborne. Each time it would just settle back down to the ground. I had to wait until early the next morning, when the cool morning air was heavier, in order to take off and come home.

Were there a lot of crashes during flight training?

Not a great deal, no. There was a periodic crash. Quite frequently a crash occurred because a pilot got in a spin. In those days the spin was referred to as the deadly spin, and it wasn't until I was fairly well along in my instruction that the spin became part of our training. You learned that as the plane begins to spin you push the nose over, pick up speed, and come out. I can remember discussing the pros and cons of this, wondering why this was. Before this the tendency was to pull the nose up, which caused the plane to stall, and with a single-engine plane that could be fatal.

Isn't it sort of astounding, in retrospect, that so may people took off with so little technical knowledge of what they were doing?

It was a good and a bad thing. It separated the sheep from the goats real quickly, but it seriously penalized the chap who might be a superb pilot but was a slow learner. Quite a large percentage of cadets were washed out in those days. Some quit of their own volition; others left because the instructor felt they were not sufficiently apt. It was pretty much of an instinctive thing.

After you were commissioned as a lieutenant in March 1918, and became an instructor yourself, what did you look for in your students?

You look for a chap who has good eyesight, who has fast reactions, who has a good sense of balance, but most important, you look for someone who really loves to fly. It would be very difficult to make a good pilot out of a chap who hated it. We always incline to do best those things that we enjoy doing. Another thing you look for is a pilot who can learn his limitations. A poor pilot is not necessarily a dangerous pilot as long as he remains within his limitations. And you find your limits in the air, by getting closer and closer and closer and sometimes going beyond them and still getting out of it. If you go beyond and don't get out of it, you haven't learned your limitations, because you are dead.

ON APRIL 18, 1942,
Lt. Col. Jimmy
DOOLITTLE
LED A FLIGHT OF 16
B-25's FROM THE DECK
OF THE U.S.S. HORNET
TO BOMB TOKYO —
THE FIRST AMERICAN
AIR RAID ON JAPAN
IN WORLD WAR II.

THE ACTUAL DAMAGE TO
THE JAPANESE WAS MINOR,
BUT THE LIFT IT GAVE A
BELEAGUERED AMERICA
WAS INCALCULABLE!

I understand that you got into trouble with one of your early commanding officers.

Ah, Colonel [Harvey] Burwell, at Ream Field, near San Diego. We later became great friends, but at that time I'm afraid I was a bit of a problem to him. His friend Cecil B. De Mille came down to Ream Field one day and took some movies, one of which showed a Jenny landing with me sitting underneath on the spreader bar between the wheels. When De Mille showed the film to Burwell, the colonel was furious and grounded me for a month and made me the permanent officer of the day for the base. This made me a bit perturbed, to say the least, so I conceived the idea of luring the colonel into an exhibition boxing match with me for the entertainment of the troops. I talked our physical director, [Charles] "Doc" Barrett, into suggesting such a match to the colonel. But Burwell correctly guessed what I was up to. The colonel had boxed at West Point, so he told Barrett that he would box *both* of us, one at a time. Doc, who wasn't much of a boxer, decided it wasn't such a good idea after all.

Wasn't it dangerous riding on the landing gear?

No. The Jenny was a very slow airplane, and I was an acrobat, a tumbler, so it was no problem at all riding on the spreader bar or even climbing out on the wing. You see, everything I ever did in aviation I practiced and practiced and practiced. As a result I was able to do things that appeared rather hazardous to someone who hadn't done them. Let me give you a better example of this. When I was the Army's chief test pilot at old McCook Field [at Dayton, Ohio] in 1927, I practiced flying the route from Dayton to Moundsville, time after time, until I had memorized every windmill, every telephone pole, every silo, and every farmhouse, so that I could fly under weather conditions where other pilots, much better pilots than I, could not fly. Yet when my commanding officer heard about it, he grounded me for being in his opinion too irresponsible, and I lost the job I enjoyed more than any job I ever had. A few years later I was flying from Cleveland to New York in bad weather. In those days they had revolving beacons every ten miles. Well, I missed a beacon, and finally I saw a light in the window of a farm-house on a hill, so I retraced my course to the previous beacon and made an emergency landing. Soon a farmer came along and informed me that the next beacon was out of order, which explained why I had missed it. Then the farmer said, by the way, the mail just went over. It was a matter of pride to me then not to let anyone fly when I couldn't, but I realized that the reason the mail had gone through was that the pilot knew that terrain so well, just as I had known the area around McCook Field. So I stayed on the ground, and I have always thought that perhaps that was the day I became a good flier, because that day I learned my limitations.

In your opinion what effect did World War I have on aviation?

The war had a dual effect. It did greatly develop our capability to conceive and manufacture airplanes, although as far as I knew, our American aircraft never saw any combat in Europe. Our pilots all flew foreign aircraft. However, by the end of the war we had a huge inventory of DH's and Jennies that lasted over a decade, and this actually stifled the manufacture of better aircraft. I can remember when we flew on the Mexican border patrol, we didn't feel the least bit badly when we would crack up an airplane. As a matter of fact, we were rather elated. because we felt we could never get better airplanes until we got rid of what we had.

By the way, how did you and your fellow pilots manage to walk away from so many crashes?

Much of it was because the airplanes were so much slower. Another part was because we crashed enough to learn *how* to crash. For instance, one night in 1929 I got into some very bad weather flying from Buffalo to New York. It became necessary to crash, so I found an opening in the clouds over a park in New Jersey and then picked a strong tree to hook my wing on. I did that deliberately. It's one of the things pilots learned. If you had to crash, the best thing was to get the outer part of your wing on something that would absorb the energy gradually instead of suddenly. They should be thinking about this same thing today for automobiles.

Why did you decide to stay in the army after 1918?

I had found a great deal of pleasure in flying, and I enjoyed the military life. Had there been commercial airlanes at that time ... no, I still would have enjoyed the military life. The boys who got out of the service and became barnstormers, they had a rather precarious existence, both from the point of view of eating and living.

What was it like in the Army Air Service after the war?

We were constantly trying to think of something to do to keep busy and, hopefully, to enhance the public's interest in aviation. The senior people particularly Billy Mitchell, understood this very clearly. We junior people didn't understand it as well, but we were anxious to participate. Any time we could get an airplane to fly someplace, well, we were for that. I remember right after the war I convinced Colonel Burwell to let three of us fly from San Diego to Washington. We didn't get very far before two of the planes cracked up. When I reported this to Burwell by telephone, he called me a Chinese ace—in those days Chinese aces were pilots who cracked up their own airplanes—and told me to come back immediately. On the way I had to make an emergency landing and flipped my plane over in a soft field. I undid my safety belt and, as

I fell out, ripped off the seat of my pants. I got the plane fixed and flew back. When I landed, a mechanic told me to report at once to the colonel. I went to his office, covered with oil, and he gave me the bawling out I was eminently entitled to. When he finished, I saluted briskly and did an about-face, exposing considerable bare posterior, which caused him to think that this was my indirect way of expressing *my* opinion. So he bawled me out again, with flourishes. I recall he said something to the effect that I couldn't even keep my ass in my pants.

Weren't you involved in Billy Mitchell's historic sinking of the captured German battleships?

You mean the *Frankfurt* and the *Ostfriesland?* Those were attacked by the Martin bombers. I was part of a squadron of DH-4's that Mitchell also assembled at Langley Field [Virginia] in 1921, and our target consisted of some smaller ships and, I think, a submarine. The mission was all part of Mitchell's belief that aviation should be a separate branch of the service. There was some merit in this concept, because the traditionalists in the army and the navy were opposed to the idea of airplanes becoming an offensive weapon. So when Mitchell finally got permission to conduct his experiment against those surplus ships, he set out to carefully train the pilots who would fly the mission. Mitchell even had special bombs made by the ordnance department, an important part of the whole operation that is sometimes overlooked.

Did you sink the targets that were assigned to your flight?

It's difficult to say. You went out and dropped your bombs and went on your way. Somebody else was scoring the results. I do have a very clear recollection of bombing an Italian battleship, the *Roma*, during World War II, and that was a very distressing mission. The *Roma* and two cruisers were anchored in the harbor at Spezia — this was before Italy came over to our side — and it was with considerable difficulty that I got permission to attack her. I put a group of B-25's on each of these ships, just to be sure, and we missed. We attacked with a hundred forty-seven bombers, and none of the three ships was sunk. It was very embarrassing. It was the only completely unsuccessful mission I led in the entire war. A little later, after Italy had switched sides, a single German light bomber with one controlled bomb came over and sank the *Roma*. I'm just glad General Mitchell wasn't around.

Billy Mitchell was an unusual man, wasn't he?

A very colorful man. I got to know him quite well. One of the busiest days I ever spent was acting as his aide on one of his missions. I was a youngster in my twenties, and I found it difficult to keep up with him. He was a man of prodigious energy. He was a good flier, too. He flew practically everything. But in the end I think that the methods he used to advance the cause of aviation probably delayed the development of air power, as well as destroyed him. I'm reminded of the two Chinese woodcuts that depict the wind blowing over a bamboo and an oak. When the wind stops, the bamboo comes back up; the oak doesn't. I'm inclined to think that if there had been a little more bamboo in Billy Mitchell, he might have achieved more than he did.

Do you agree with the outcome of Mitchell's court-martial?

My feeling then and now is that Mitchell was right about the basic principles involved. He was ahead of his time in that. His concept of the 1921 bombing maneuvers and their execution was absolutely brilliant. But when he defied authority to the extent of bringing a court-martial on himself, I think he possibly went too far. I feel the same way about that [1951] altercation between President Truman and General MacArthur. MacArthur was absolutely right, but he defied the President, and Truman was absolutely right in kicking him out, because you cannot have a lack of discipline in the high echelons of command and expect to maintain discipline in the lower echelons.

Wasn't MacArthur one of the officers on Mitchell's court-martial board?

Yes, and I remember visiting MacArthur in his Tokyo headquarters just before Truman relieved him. I had a very pleasant couple of hours with him — he was a fascinating talker, a very knowledgeable man, a great leader, a great American — and somehow our discussion got around to Billy Mitchell. And MacArthur said, "Did you know that there was one member of that board and one member only who voted to acquit Billy Mitchell?" MacArthur said, "I was that one member." The voting, of course, on the court-martial board was never announced, although in the inner circle we had heard that MacArthur had voted for acquittal. But I did not know for sure until I heard it from his mouth.

In September 1922, you made the first one-stop cross-country flight in less than twenty-four hours. How did this come about?

Well, all of us in aviation were interested in advancing two things. One was aviation, and the other was ourselves. A chap named Alex Pierson came up with this idea, but he was forced down in Mexico and abandoned the flight. So I took a DH-4, had additional tankage installed, and then borrowed a brand new instrument, a bank and turn indicator, which was still undergoing testing at Dayton, Ohio. I was all set to take off one night from Pablo Beach, outside Jacksonville. Florida, but after I gave it the gun, my left wheel hit a soft spot in the sand. The next thing I knew, I was upside down in the water. I unbuttoned my safety belt, but as I fell out my helmet and goggles went down over my eyes and shut off my nose. I thought I was underwater. So I began climbing up the fuselage. I got clean on top of it before I realized I was in knee-deep water, and it was my goggles that were shutting off my breathing. The crowd that had gathered to see me take off thought my desperate antics were pretty funny. I was very embarrassed. I immediately requested permission to have the airplane repaired, and then I flew to San Diego, with a refueling stop at Kelly Field in Texas, in twenty-one hours and nineteen minutes. Kelly and Macready made the same flight, this time non-stop, the following year; and that flight, of course, was a much more difficult and justifiably more heralded flight.

You were awarded your first Distinguished Flying Cross for your flight. And then the army sent you to MIT to study aeronautical engineering. Would you tell me about that?

There were six pilots selected for that training, and basically the idea was to get more rapport between the aeronautical engineer and the pilot. In those days there was a general feeling among pilots that the aeronautical engineers were not quite as competent as they should be. The engineers, on the other hand, felt that the pilots were all a little touched in the head or they wouldn't be pilots in the first place. So we were the first group really to try to bridge that gap.

Do you think you were successful?

Well, the thesis I wrote for my master's degree — which became known as "N.A.C.A. Report No. 203: Accelerations in Flight" — was published in every technical language in the world. The report was a rather unique thing at the time, because I took an airplane up to failure. The last pullout I made went up, I think, to 7.8 G's, and the wing failed but did not come off. This gave us a chance for the first time to check the strength calculations against the actual loads imposed in flight. I was also able to supply some scientific information about the actual effects of prolonged acceleration on the human body. Not too much was known then about G [gravity] forces. A lot of pilots were under the impression that when they blacked out, the only faculty they lost was their sight. My experiments indicated that sight was the last faculty to be lost under those conditions. It was a very useful piece of work.

What was the subject of your doctoral dissertation?

I studied the wind-velocity gradient and its effect on flying characteristics. I was very interested to find that among the most experienced test pilots there was a great difference of opinion on this subject. They all insisted, for instance, that they could always "feel" the wind direction. Some of them also claimed that they could always tell the attitude of their ship, whether they could see the horizon or not. And it was true then that most of the airplanes were so stable that if you just let them go, they would level off. Because of that inherent stability in the airplane, many pilots felt that they could sense their ship's attitude through something real lucky in their butts. My experiments, however, proved that they were quite wrong. My tests showed that when you got far enough from the ground in a steady wind, it made no difference whether you turned into or out of the wind, your reactions were the same as if you were in a dead calm. . . .

Would you say that your theoretical studies at MIT gave you greater confidence as a flier?

It is true that, particularly after my work on acceleration, I had a much better idea of the stresses that an airplane could be expected to stand up under. But I believe it is equally true that within my personal limitations as a pilot I have always carefully calculated every risk I ever took. I constantly practiced, and I only did those things that I could do relatively safely.

Even from the ground such things as the formation stunt-flying you did just after World War I looked extremely dangerous. Incidentally, were you the first to stunt in formation?

I don't think so. I was certainly among the first, but I don't think I conceived the idea. It's sort of an evolutionary thing. Two folks get together and do some stunting, then three, and then, my goodness, five would be dandy. So you end up with five planes. I did later pioneer something that was called the apron-string event. At McCook Field five of us tied our airplanes together with fifteen-foot ribbons, took off, went through some maneuvers, stunted, and landed with the planes still tied together. We did it at several air shows, and as far as I know, that had never been done before.

In a stunt like that you have to have a great deal of confidence in the lead pilot, don't you?

You have to have a great deal of confidence in the people who are behind you, too!

"Behind you" was where all the other American, British, and Italian pilots finished in the 1925 Jacques Schneider Maritime Cup Race. Would you tell me about it?

I consider this a very important event in my life, and I'd like to explain something about these early airplane races. A lot of good came out of them, especially in aircraft design, just as car racing, for instance, led to the tremendous improvement in automobile tires. But it was quite expensive, and neither the army nor the navy could afford to build these new planes and engines. So what happened was the army and the navy got together, and each put up $250,000 for the Curtiss Company to build four new planes, which were the most modern in the world in 1925. One of these air frames was statically tested. That left three airplanes and one spare engine. As was customary, the navy got two and the army one. I never quite understood that, but that was the way it always worked out. At any rate, I had the benefit of some aeronautical engineering training, and I was able to change to a propeller with a slightly different pitch and thus pull optimum speed out of the engine.

You didn't have adjustable props in those days?

You couldn't feather wooden propellers. It wasn't until much later that metal came in, and it wasn't for years after that that they became adjustable. Incidentally, the first adjustable prop was not made to pull different power from the engine but rather as a reversible prop, so that you could stop more quickly upon landing. I believe it was in 1922 that Sandy Fairchild, who later became vice chief of the air staff, tested the first reversible-pitch propeller. But it reversed on takeoff and dropped Sandy into the river, and that set back the development of the controllable-pitch propeller about two decades. There was a great lack of interest in something that took control of the plane when you didn't want it to.

In the 1925 Schneider Cup your winning speed of 232.57 miles an hour around the pylons on Chesapeake Bay shattered the old record of 177 miles an hour. You didn't accomplish this just by changing the propeller pitch, did you?

First of all, I was the army pilot selected to test the new Curtiss racing plane, so I became quite familiar with the aircraft. Then the Navy was most cooperative about letting me practice in various seaplanes before the race. I was able to develop a useful system, which was to come in a little above the pylon and lose some altitude as I whipped around it. You see, if you tended to climb in a turn, you lost speed rapidly. So my stock-in-trade in racing was to come out of the pylon turn losing a little altitude but not much speed. Then on the next leg I could gain that fifty feet more or less, without having to lose any speed en route.

Was the navy embarrassed about an army pilot winning a seaplane race?

They weren't exactly elated about my victory, and after I got back to McCook Field, my colleagues gave me a parade through Dayton and made me ride in a rowboat and called me Admiral Doolittle. But let me tell you something interesting about the Schneider Cup. This was an international competition, and it was agreed that which-ever nation won the race three times in a row would permanently retire the cup. Now,

I won in 1925. Two years before [Lieutenant David] Rittenhouse of the navy won it. But there was no race in 1924 because the British airplane had been wrecked loading it onto a ship. All the Americans would have had to do in 1924 was to fly around the pylons at any speed to win the race, and with my victory the following year the United States would have retired the cup. But because the British were unable to compete, the Americans, out of good sportsmanship, did not choose to race in 1924. So what happened was that the Italians won the race in 1926, and the English got tremendously interested, and I believe a considerable amount of money was put up by an English woman for research. The result was that the British came back and won the race three times and captured the Schneider Cup. But also out of that same research came the engines and air frames and concepts that went into the Hurricanes and Spitfires, without which the whole Second World War might have had a very different conclusion.

Are you saying that because the Americans were good sports and didn't fly the 1924 race, this country actually contributed to the development of British aviation?

I feel strongly that the British Schneider Cup racing planes had a profound effect on World War II. And curiously enough, the year the British won their third race and took the cup, there was no other competition!

What was your next assignment after the Schneider Cup?

Curtiss requested that the army give me a leave of absence so that I could demonstrate the P-1 fighter that the company was trying to sell to certain South American governments. In Chile, the night before we were to demonstrate the airplanes—Germany, England, and Italy were also represented—I was at a party in an officers' club in Santiago, and the talk got around to the movies and Douglas Fairbanks. Well, one of the Chileans, who of course didn't know that I was a tumbler, asked me what I thought about Fairbanks's acrobatics, and I told him that all American kids were trained to do those stunts. I did a few elemental gymnastic stunts for them, and then I went into a handstand on the window sill. But as I lowered my legs out parallel the sill collapsed, and I fell about twenty feet to the ground, breaking both my ankles. I didn't feel that I could let down the army or Curtiss, so after the doctor put casts on my legs, I had special clips attached to the P-1's rudder bars to hold my feet on the controls. I think what really sealed the sales contract with Chile was a little performance of mock aerial combat I put on the next day with Ernest von Schonabeck, who had been an ace in the Richthofen Squadron.

Did you go into the hospital after the flight demonstration?

Eventually I did. I spent six months in Walter Reed General Hospital when I got back to the States, but that was after we had demonstrated the P-1 in Bolivia and Argentina.

When I took the airplane to Buenos Aires, I became the first American to fly across the Andes. The flight was made without a parachute, because with my feet strapped to the controls I couldn't have jumped anyway.

When did parachutes come into general use?

We did not begin to use parachutes until the early twenties. We didn't have them during the first war. The helmet and goggles were the insignia of our profession, and, you know, a parachute just didn't feel like part of our uniform. But one day at McCook Field a group of us were watching Harold Harris when his airplane disintegrated in the air. Harris became the first man to save his life with a parachute, and it made instant believers out of those of us on the ground. I'm not sure, but I think that after the Harris episode I wore one all the time.

Did you ever have occasion to use a parachute?

The parachute saved my life three times. Once in 1929, once in 1932, and the last time was after the Tokyo raid and I had to bail out over China.

What happened in 1929?

At the time I was stationed at Mitchel Field on Long Island, carrying out the Guggenheim blind-flying experiments, and General Patrick called and asked me to perform some acrobatics at an air show in Cleveland. I went out ahead of time to practice with a little Curtiss Hawk, a fine little biplane that I was well acquainted with. This particular plane had been modified—the nose had been streamlined by moving the radiator to the upper wing—and as a result it probably dove faster than the conventional Hawk. I had recently flown the first outside loop, and General Patrick had issued a directive saying that nobody was to do any more outside loops. But the general had not said that you couldn't push a plane under and turn it out. I thought this would be a rather spectacular stunt, but when I tried it in practice, the wings folded up. I just unbuckled the safety belt and was thrown out. I parachuted down and immediately got another airplane, finished my practicing, and did the air show that afternoon.

Is it important after an experience like that to get right back into the air, sort of like remounting a horse that's thrown you?

I am not a very timid type. It's very important to some people, but not to me. I have a simple philosophy: worry about those things you can fix. If you can't fix it, don't worry about it; accept it and do the best you can. . . .

This period of your career—1917 to 1930—encompasses the frontier days of aviation. Do you consider these years the most exciting in the history of aviation?

That's a hard question to answer. I will say that in those days the pilot was very important, and his skill in manipulating the airplanes, which were not as reliable as they are today, was very important indeed. The airplanes today are mechanized to such a degree that the pilot no longer depends on the seat of his pants to the extent that he did in the early days. What has happened to aviation has happened to almost everything else. The day of the rugged individualist, the day of the inventor, is almost over. The Ben Franklins and Henry Fords are pretty much a thing of the past. It has just become too complicated. Everything now is a team operation, and if a truly new concept is developed, it means that there will be a large number of people knowledgeable in various scientific disciplines involved. And this requires a different philosophical outlook. I cannot see, for instance, how we could ever have another Lindbergh. Things have changed too much for that sort of competence to be rewarded the way it justifiably was. Still, I think that aviation will continue to develop, and each era will be interesting. But interesting in different ways.

Looking back to those early days, does it ever amaze you that so many of the pioneer aviators, yourself included, are still alive?

It amazes me that so many of them are gone.

This poem is the last page of Amelia Earhart's *Last Flight*, a book issued in 1937 after her disappearance in the Pacific Ocean. She and her navigator, Fred Noonan, were attempting to circle the globe, it was said, and had traversed the skies from California to Brazil, Africa, India, and New Guinea. Then, somewhere between New Guinea and Howland Island, they vanished. Not until nearly thirty years had passed did the truth emerge. Earhart and Noonan were on a secret mission: to observe the Japanese military facility on the island of Truk, particularly its number of airfields and fleet-servicing capability. The pair became confused in bad weather, however, and went down on Mili Atoll in the Japanese-held Marshall Islands. Earhart and Noonan were brought for interrogation to Saipan, where they met their deaths.

AMELIA EARHART

Courage

Courage is the price that life exacts for granting peace.
The soul that knows it not, knows no release
From little things;

Knows not the livid loneliness of fear
Nor mountain heights, where bitter joy can hear
The sound of wings.

How can life grant us boon of living, compensate
For dull gray ugliness and pregnant hate
Unless we dare

The soul's dominion? Each time we make a choice, we pay
With courage to behold resistless day
And count it fair.

René Fonck and Georges Guynemer, France's two great fighter pilots of World War I, were of differing temperaments and not truly friends. Still, when Guynemer went down after fifty-four victories, Fonck mourned with France and took up where his fallen comrade left off. Not only did he avenge Guynemer, but he surpassed his total with seventy-five and at war's end emerged as the Allies' ace of aces.

RENE FONCK

The Avenger

The 11th of September, 1917, must be remembered as a black day.

On that day the air corps lost its most glorious hero. In prose and in verse, others have praised his exploits, expressed all the sorrows of the nation. Parliament voted him the Honor of the Pantheon; but the best mark of admiration and of regret is the one which has remained engraved in the hearts of his fighting companions, who all became his avengers and among whom many fell in this endeavor.

Guynemer left on patrol at the break of day. The first rays of sun, showing through the mist, sparkled on the foliage already turning brown because of the coolness of the nights.

While the mechanics were scurrying about his plane, the pilot, a bit nervous, as if a hidden fear was haunting him, was pacing feverishly, like a caged lion.

"Guynemer is in a bad humor this morning," observed someone, but none of us thought, at the moment, of attaching any importance to the remark.

Finally the motor started up and our comrade climbed quickly into the cockpit, the propeller whirred and like a great bird, the plane took off.

Lieutenant Bozon-Verduras left at the same time. What happened after that? . . . We never learned the facts precisely. We only knew that they flew together and helped each other in several engagements.

Lieutenant Bozon, separated from his companion by an unexpected pursuit, lost sight of him and returned alone to the field.

It was not the first time that Guynemer had returned late, therefore no one even thought of worrying, but as the hours passed, his absence became the subject of all conversation. Commandant Brocard no longer hid his concern. Information was requested by telephone from the front lines and we were soon notified that a French plane had been seen falling behind the German lines.

Those who had observed the plane falling asserted, however, that it was not in flames and that the pilot might not be injured.

Two or three days went by and Guynemer's return became more and more doubtful. We hoped nevertheless that he had made out all right and that escape or peace would bring him back to us, but an enemy newspaper deprived us even of this hope. It published, along with the news of the death of our national hero, the name of the man who had shot him down . . . Wissemann.

I was near the hangars when Commandant Brocard brought me the sad confirmation. Instantly I had my fighter plane brought out, determined to seek an escape from my deep sorrow, in the air.

Ten minutes after my departure, I spotted an airplane off in the distance. I realized immediately that it was a two-seater observation plane. Absorbed in their work, the two passengers had not seen me coming. As usual I climbed very high in order to dive on the enemy. This tactic, common to birds of prey and which is instinctive to them, has always seemed the very best to me.

I surprised them in an attitude of complete security, the pilot with his hands on the throttle and the observer in the process of taking pictures, leaning out over the fuselage.

I swooped down but waited until I was a few yards away before opening fire. With my eyes glued to the sight, I could see all details growing rapidly in size. Aiming directly for the center of the plane, my bullets swept across the motor and crew.

It wasn't long before I saw the results. Killed, no doubt, by the very first burst, the pilot must have slumped in his seat, jamming the controls in his agony, for the plane immediately began to spin. While I was avoiding a collison by making a rapid turn, I saw the live body of the observer fall from the upside-down plane. For a brief moment he had tried to remain in his seat. He passed by, his arms frantically clutching at the emptiness, a few yards from my left wing. . . . I will never forget this sight. . . .

But the incendiary bullets had done their job, and the plane like a gigantic torch dropped at full speed toward the earth, preceded by the body of the observer.

A few days later our group left Flanders.

There, foggy regions, where the never ending rain and fog are likely to keep a squadron such as the "Cigognes" ("Storks") on the ground for an entire winter. At that time the retention of such a unit in this area would have been a real waste of power.

During autumn, we were hardly able to claim even a few victories for ourselves, and for my last success I brought down a large two-seater reconnaissance plane. On September 30, 1917, I was up about 12,000 feet with several comrades, when I saw this bold fellow making a turn below us. He continued his inspection calmly as if his position was completely free of danger.

Immediately I opened my throttle and my Spad leaped forward. The enemy plane didn't seem to be worried, but upon approaching I saw the gunner at his combat post. The maneuver required close precision, but I was an old hand at this game. While closing on him, I made my plane rise and fall in a flitting manner similar to that which permits a butterfly to evade his enemies, and my tactics evidently seemed to disconcert him.

I do not subscribe to the historic chivalrous remark made by Fontency. That officer who was in command at the time thought it necessary to say: "Englishmen, you may fire first." This is a very archaic method. If the English had had machine guns at their disposal at that time, not one single Frenchman would probably have survived to report these gallant words. I believe that it is necessary to adopt a middle course, not to waste your ammunition and to fire your bullets at the very instant where they have a chance to reach their target.

Therefore, without response, I accepted the fire of the Boche three times. But when my turn came I fired my machine gun and had the satisfaction of seeing the enemy plane turn. I made a few feints and rapidly broke off, having succeeded in placing myself under my adversaries' control surfaces. I hit the pilot and the gunner almost simultaneously.

From above, I watched their fall, perhaps 3,000 feet below me, one of the wings broke loose and the men were hurled out of their bucket seats. I always experienced a certain compassion for my victims at such a sight, in spite of the satisfaction, somewhat ani-mallike, of having saved my own skin, and the patriotic joy of victory.

I would often prefer to spare their lives, especially when they have fought coura-geously, but as I believe I've already said, in air battles there is generally no alternative but victory or death, and it is rarely possible to show mercy without betraying the interests of one's country.

Thus, these last Germans brought down had been admirable; neither the pilot nor the gunner had lost his composure for an instant, and upon seeing me approach from a distance, instead of abandoning their task of reconnaissance and going back to their lines with throttle wide open, they had waited for me solidly accepting battle without flinching. Immediately upon returning to the field I took a car to go and look at my victims and examine the remains of the plane to find out if they had any new technical devices. Some officers had preceded me and the first news that they announced was that one of the bodies carried identification in the name of Wissemann — the very one that the German papers had publicized as the pilot who shot down Guynemer.

In 1783 the United States diplomatic representative to France was Benjamin Franklin. Fortunately for posterity, he witnessed three of the celebrated balloon flights of that *annus mirabilis*, and he recorded what he saw. The first aeronautic letter describes the August 27 hydrogen-balloon experiment of J. A. C. Charles. (As the balloon rose, a fellow spectator is said to have asked Franklin, "What good is it?" The classic reply: "What good is a newborn baby?") The second letter describes the November 21 flight of Pilâtre de Rozier and the Marquis d'Arlandes (whose own account appears earlier in this volume). And the third letter details the flight, only ten days later, of Professor Charles and Noël Robert in a hydrogen balloon.

BENJAMIN FRANKLIN

Balloons Over Paris

PASSY, AUGUST 30, 1783

On Wednesday, the 27th instant, the new aerostatic experiment, invented by Messrs. Montgolfier of Annonay, was repeated by M. Charles, Professor of Experimental Philosophy at Paris.

A hollow globe twelve feet diameter was formed of what is called in England oiled silk, here taffetas gommé, the silk being impregnated with a solution of gum elastic in linseed oil, as he said. The parts were sewed together while wet with the gum, and some of it was afterwards passed over the seam, to render it as tight as possible.

It was afterwards filled with inflammable air that is produced by pouring oil of vitriol upon fillings of iron, when it was found to have a tendency upwards so strong as to be capable of lifting a weight of 39 pounds, exclusive of its own weight which was 25 pounds and the weight of the air contained.

It was brought early in the morning to the Champ de Mars, a field in which reviews are sometimes made, lying between the military school and the river. There it was held down by a cord till five in the afternoon, when it was to let loose. Care was taken before the hour to replace what portion had been lost, of the inflammable air, or of its force, by injecting more.

It is supposed that not less than 50,000 people were assembled to see the experiment, the Champ de Mars being surrounded by multitudes, and vast numbers on the opposite side of the river.

At five o'clock notice was given to the spectators by the firing of two cannon, that the cord was about to be cut. And presently the globe was seen to rise, and that as fast as a body of 12 feet diameter, with a force only of 39 pounds, could be supposed to remove the resisting air out of its way. There was some wind, but not very strong. A little rain had wet it, so that it shone, and made an agreeable appearance. It diminished in apparent magnitude as it rose, till it entered the clouds, when it seemed to me scarce bigger than an orange, and soon after became invisible, the clouds concealing it.

The multitude separated, all well satisfied and delighted with the success of the experiment, and amusing one another with discourses of the various uses it may possibly be applied to, among which many were very extravagant. But possibly it may pave the way to some discoveries in natural philosophy of which at present we have no conception.

A note secured from the weather had been affixed to the globe, signifying the time and place of its departure, and praying those who might happen to find it, to send an account of its state to certain persons at Paris. No news was learned of it till the next day, when information was received that it fell a little after six o'clock, at Gonesse, a place about four leagues distance, and that it was rent open, and some say had ice in it. It is supposed to have burst by the elasticity of the contained air when no longer compressed by so heavy an atmosphere.

One of 38 feet diameter is preparing by M. Montgolfier himself, at the expense of the Academy, which is to go up in a few days. I am told it is constructed of linen and paper, and is to be filled with different air, not yet made public, but cheaper than that produced by the oil of vitriol, of which 200 Paris pints were consumed in filling the other.

It is said that for some days after its being filled the ball was found to lose an eighth part of its force of levity in twenty-four hours; whether this was from imperfection in the tightness of the ball, or a change in the nature of the air, experiments may easily discover. . . .

M. Montgolfier's air to fill the globe has hitherto been kept secret; some suppose it to be only common air heated by passing through the flame of burning straw, and thereby extremely rarefied. If so, its levity will soon be diminished by condensation, when it comes into the cooler region above. . . .

P. S. I just now learned that some observers say, the ball was 150 seconds in rising, from the cutting of the cord till hid in the clouds; that its height was then about 500 toises [3,200 feet], but, being moved out of the perpendicular by the wind, it had made a slant so as to form a triangle, whose base on the earth was about 200 toises. It is said the country people who saw it fall were frightened, conceived from its bounding a little, when it touched the ground, that there was some living animal in it, and attacked with stones and knives, so that it was much mangled; but it is now brought to town and will be repaired.

The great one of M. Montgolfier is to go up, as is said, from Versailles, in about eight or ten days. It is not a globe but of a different form, more convenient for penetrating the air.

It contains 50,000 cubic feet, and is supposed to have force of levity equal to 1,500 pounds weight. A philosopher here, M. Pilâtre du Rozier, has seriously applied to the academy for leave to go up with it, in order to make some experiments. He was complimented on his zeal and courage for the promotion of science, but advised to wait till the management of these balls was made by experience more certain and safe. They say the filling of it in Montgolfier's way will not cost more than half a crown. One is talked of to be 110 feet diameter. Several gentlemen have ordered small ones to be made for their amusement. One has ordered four of 15 feet diameter each; I know not with what purpose; but such is the present enthusiasm for promoting and improving this discovery, that probably we shall soon make considerable progress in the art of constructing and using the machines.

Among the pleasantries conversation produces on this subject, some suppose flying to be now invented, and that since men may be supported in the air, nothing is wanted but some light handy instrument to give and direct motion. Some think progressive motion on the earth may be advanced by it, and that a running footman or a horse slung and suspended under such a globe so as to have no more of weight pressing the earth with their feet than perhaps eight or ten pounds, might with a fair wind run in a straight line across countries as fast as that wind, and over hedges, ditches, and even waters. It has been even fancied that in time people will keep such globes anchored in the air, to which by pulleys they may draw up game to be preserved in the cool and water to be frozen when ice is wanted. And that to get money, it will be contrived to give people an extensive view of the country, by running them up in an elbow chair a mile high for a guinea, etc., etc.

PASSY, NOVEMBER 22, 1783

Enclosed is a copy of the *Procès-verbal* taken of the experiment yesterday in the garden of the Queen's Palace la Muette, where the Dauphin now resides, which being near my house I was present. This paper was drawn up hastily, and may in some places appear to you obscure; therefore I shall add a few explanatory observations.

This balloon was larger than that which went up from Versailles and carried the sheep, etc. Its bottom was open, and in the middle of the opening was fixed a kind of basket grate, in which faggots and sheaves of straw were burnt. The air rarefied in passing through this flame rose in the balloon, swelled out its sides, and filled it.

The persons who were placed in the gallery made of wicker, and attached to the outside near the bottom, had each of them a port through which they could pass sheaves of straw into the grate to keep up the flame, and thereby keep the balloon full.

When it went over our heads, we could see the fire which was very considerable. As the flame slackens, the rarefied air cools and condenses, the bulk of the balloon diminishes, and it begins to descend. If those in the gallery see it likely to descend in an improper place, they can by throwing on more straw, and renewing the flame, make it rise again, and the wind carries it farther.

One of these courageous philosophers, the Marquis d'Arlandes, did me the honor to call upon me in the evening after the experiment, with Mr. Montgolfier, the very ingenious inventor. I was happy to see him safe. He informed me that they lit gently, without the least shock, and the balloon was very little damaged.

This method of filling the balloon with hot air is cheap and expeditious, and it is supposed may be sufficient for certain purposes, such as elevating an engineer to take a view of an enemy's army, works, etc., conveying intelligence into, or out of a besieged town, giving signals to distant places, or the like.

The other method of filling a balloon with permanently elastic inflammable air, and then closing it is a tedious operation, and very expensive; yet we are to have one of that kind sent up in a few days. It is a globe of 26 feet diameter. The gores that compose it are red and white silk, so that it makes a beautiful appearance. A very handsome triumphal car will be suspended to it, in which Messrs. Roberts, two brothers, very ingenious men, who have made it in concert with Mr. Charles, propose to go up. There is room in this car for a little table to be placed between them, on which they can write and keep their journal, that is, take notes of everything they observe, the state of their thermometer, barometer, hygrometer, etc., which they will have more leisure to do than the others, having no fire to take care of. They say they have a contrivance which will enable them to descend at pleasure. I know not what it is. But the expense of this machine, filling included, will exceed, it is said, 10,000 livres.

This balloon of only 26 feet diameter, being filled with air ten times lighter than common air, will carry up a greater weight than the other, which though vastly bigger, was filled with an air that could scarcely be more than twice as light. Thus the great bulk of one of these machines, with the short duration of its power, and the great expense of filling the other will prevent the inventions being of so much use as some may expect, till chemistry can invent a cheaper light air producible with more expedition.

But the emulation between the two parties running high, the improvements in the construction and management of the balloons had already made a rapid progress; and one cannot say how far it may go. A few months since the idea of witches riding through the air upon a broomstick, and that of philosophers upon a bag of smoke, would have appeared equally impossible and ridiculous.

These machines must always be subject to be driven by the winds. Perhaps mechanic art may find easy means to give them progressive motion in a calm, and to slant them a little in the wind.

I am sorry this experiment is totally neglected in England, where mechanic genius is so strong. I wish I could see the same emulation between the two nations as I see between the two parties here. Your philosophy seems to be too bashful. In this country we are not so much afraid of being laughed at. If we do a foolish thing, we are the first to laugh at it ourselves, and are almost as much pleased with a bon mot or a *chanson*, that ridicules well the disappointment of a project, as we might have been with its success. It does not seem to me a good reason to decline prosecuting a new experiment which apparently increases the power of a man over matter, till we can see to what use that power can be applied. When we have learnt to manage it, we may hope some time or other to find uses for it, as men have done for magnetism and electricity, of which the first experiments were mere matters of amusement.

This experience is by no means a trifling one. It may be attended with important consequences that no one can foresee. We should not suffer pride to prevent our progress in science.

Beings of a rank and nature far superior to ours have not disdained to amuse themselves with making and launching balloons, otherwise we should never have enjoyed the light of those glorious objects that rule our day and night, nor have had the pleasure of riding round the sun ourselves upon the balloon we now inhabit.

PASSY, DECEMBER 1, 1783

In mine of yesterday I promised to give you an account of Messrs. Charles and Roberts' experiment, which was to have been made this day, and at which I intended to be present. Being a little indisposed, and the air cool, and the ground damp, I declined going into the Garden of the Tuilleries where the balloon was placed, not knowing how long I might be obliged to wait there before it was ready to depart; and chose to stay in my carriage near the statue of Louis XV, from whence I could well see it rise, and have an extensive view of the region of air through which, as the wind sat, it was likely to pass. The morning was foggy, but about one o'clock the air became tolerably clear; to the great satisfaction of spectators, who were infinite. Notice having been given of the intended experiment several days before in the papers, so that all Paris was out, either about the Tuilleries, on the quays and bridges, in the fields, the streets, at the windows, or on the tops of houses, besides the inhabitants of all the towns and villages of the environs. Never before was a philosophical experiment so magnificently attended. Some guns were fired to give notice that the departure of the great balloon was near, and a small one was discharged which went to an amazing height, there being but little wind to make it deviate from its perpendicular course, and at length the sight of it was lost. Means were used, I am told, to prevent the great balloon's rising so high as might endanger its bursting. Several bags of sand were taken on board before the cord that held it down was cut, and the whole weight being then too much to be lifted, such a quantity was discharged as to permit its rising slowly. Thus it would sooner arrive at that region where it would be in equilibrio with the surrounding air, and by discharging more sand afterwards, it might go higher if desired. Between one and two o'clock, all eyes were gratified with seeing it rise majestically from among the trees and ascend gradually above the buildings, a most beautiful spectacle! When it was about two hundred feet high, the brave adventurers held out and waved a little white pennant, on both sides their car, to salute the spectators, who returned loud claps of applause. The wind was very little, so that the object, though moving to the northward, continued long in view; and it was a great while before the admiring people began to disperse. The persons embarked were Mr. Charles, Professor of Experimental Philosophy, and zealous promotor of that science; and one of the Messrs. Robert, the very ingenious constructors of the machine. When it arrived at its height, which I suppose might be 300 or 400 toises, it appeared to have only horizontal motion. I had a pocket glass, with which I followed it, till I lost sight first of the men, then of the car, and when I last saw the balloon, it appeared no bigger than a walnut. I write this at seven in the evening. What became of them is not yet known here. I hope they descended by daylight, so as to see and avoid falling among trees or on houses, and that the experiment was completed without any mischievous accident, which the novelty of it and the want of experience might well occasion. I am the more anxious for the event, because I am not well informed of the means provided for letting themselves gently down, and the loss of these very ingenious men would not only be a discouragement to the progress of the art, but be a sensible loss to science and society.

Tuesday morning, December 2 — I am relieved from my anxiety by hearing that the adventurers descended well near l'Isle Adam, before sunset. This place is near 7 leagues from Paris. Had the wind blown fresh, they might have gone much farther.

P.S. Tuesday evening . . . I hear farther that the travelers had perfect command of the carriage, descending as they pleased by letting some of the inflammable air escape, and rising again by discharging some sand; that they descended over a field so low as to talk with laborers in passing and mounted again to pass a hill. The little balloon falling at Vincennes shows that mounting higher it met with a current of air in a contrary direction; an observation that may be of use to future aerial voyagers.

The author and title tell it all: one of the most influential men of all time commenting upon another. From the monograph *Leonardo da Vinci: A Study in Psychosexuality*, translated from the German by A. A. Brill.

SIGMUND FREUD

Leonardo da Vinci: The Erotic Roots of Aviation

A very obscure as well as a prophetically sounding passage in his [Leonardo's] notes dealing with the flight of the bird demonstrates in the nicest way with how much affective interest he clung to the wish that he himself should be able to imitate the art of flying: "The great bird shall take his first flight from the back of his big swan, filling the world with amazement, all writings with his fame, and bring eternal glory to the nest whence he sprang." He probably hoped that he himself would sometimes be able to fly, and we know from the wish-fulfilling dreams of people what bliss one expects from the fulfillment of this hope.

But why do so many people dream of flying? Psychoanalysis answers this question by stating that to fly or to be a bird in the dream is nothing but a concealment of another wish, to the recognition of which we are led by more than one linguistic or real bridge. When the inquisitive child is told that a big bird like the stork brings the little children, when the ancients have formed the phallus winged, when the popular designation of the sexual activity of man is expressed in German by the word "to bird" *(vögeln)*, when the male member is directly called *l'uccello* (bird) by the Italians, all these facts are only small fragments from a large collection which teaches us that the wish to be able to fly signifies in the dream nothing but the longing for the ability of sexual accomplishment. This is an early infantile wish. When the grown-up recalls his childhood, it appears to him as a joyful time, in which one is happy for the moment and looks to the future without any wishes; it is for this reason that he envies children. But the children themselves, if they could inform us about it, would probably give us different reports. It seems that childhood is not that blissful idyl into which we later distort it, that, on the contrary, children are lashed through the years of childhood by the wish to become big, and to imitate the grown-ups. This wish instigates all their playing. If in the course of their sexual investigation children feel that the grown-up knows something wonderful concerning the mysterious and yet so important realm that they are prohibited from knowing or doing, they are seized with a violent wish to know it, and dream of it in the form of flying, or prepare this disguise of the wish for their later flying dreams. Thus aviation, which has attained its aim in our times, has also its infantile erotic roots.

By admitting that he entertained a special personal relation to the problem of flying since his childhood, Leonardo confirms what we must assume from the investigation of children of our own times, namely, that his childhood investigation was directed to sexual matters.

If an editor is compiling an anthology of aviation writing, he must have *something* by Ernest K. Gann. But what? How to choose from *Fate Is the Hunter, The High and the Mighty, In the Company of Eagles,* or *Blaze of Noon?* The decision: match this most popular aviation writer with the most important plane in the history of air transportation. The Douglas DC-3 was the first aircraft to turn a profit solely by hauling passengers. It was an overbuilt, incredibly sturdy plane that, in its military version, the C-47, proliferated by the thousands. After World War II the C-47 was frequently sold as surplus and converted to commercial use, returning to service with some small airline. This paean to a great plane is from *Ernest K. Gann's Flying Circus.*

ERNEST K. GANN

The Masterpiece

A time would come when a great host of people would claim the honors for inspiring the Douglas Sleeper Transport (DST), an aircraft which soon emerged from its beautiful cocoon and became the legendary DC-3. As in all human affairs every masterpiece acquires a multitude of latter-day sponsors, but in the conception of the DST two men stand in a fort of solid truth. They are the remarkable and much beloved C. R. Smith, who then headed American Airlines, and William Littlewood, the line's chief engineer. An extraordinarily enterprising pair, the pithy, venturesome sort of men who originally forged the airlines of the United States, they persuaded Donald Douglas to expand his already successful DC-2s into a "sleeper" version. The eventual result was an aircraft which resembled the DC-2 only when viewed from a great distance.

If the DST was justly rated as the queen of all contemporary aircraft then her aerodynamically identical sister, the DC-3, must be considered the simple working member of the family, short on glamour while blessed with productivity and hence assured of a very much longer life. As in fairy tales it was the humble chargirl who became famous and exerted a powerful effect upon world history.

The DC-3 was and is unique, for no other flying machine has been a part of the international scene and action so many years, cruised every sky known to mankind, been so ubiquitous, admired, cherished, glamorized, known the touch of so many different pilot nationals, and sparked so many maudlin tributes. It was without question the most all-up successful aircraft ever built and even in this jet age it seems likely the surviving Douglas DC-3s, full-blood sisters to the long forgotten DSTs, may fly about their business forever. One DC-3 flown by North Central Airlines logged 84,000 hours air time as of June 1966, more than any other aircraft in the history of flight.

A late October afternoon in 1936. Newark Airport in New Jersey, the only airline terminus for the entire metropolitan area of New York. The world is uneasy—everywhere. Hitler has reoccupied the Rhineland in defiance of the Locarno Pact. The new and very bloody civil war which erupted in July has already torn Spain to pieces and the defunct League of Nations has abandoned Ethiopia to the Italians. Yet for Americans life is not at all bad. Rib roast sells for only 31 cents a pound, and track star Jesse Owens has won four gold medals in the Berlin Olympic Games. Among other amenities, well-funded people can now avoid the interminable rail journey from coast to coast by booking passage in one of American Airlines' glistening DSTs.

The operation is now routine, which means that the much-publicized "Mercury" flight to Los Angeles is supposed to depart Newark at 5:10 and touch down on Glendale airport at 8:50 tomorrow morning—God and the elements willing. The average has been reasonably good, but the prevailing westerly winds across the North American continent often make the public relations people wish the schedule was printed on elastic.

For the fourteen passengers there are certain consolations to curb their impatience and the two movie celebrites who have paid more than the $150 standard fare for the privilege of occupying the private compartment known as the "Sky Room" are duly and regularly soothed by the not-so-subtle attentions of the stewardess. Once in the air, all souls aboard with the exception of crew will be served cocktails on the house followed by a fine steak dinner. Then Captain Dodson will send back his written flying report to be passed among his guests. It will be signed by the copilot, who will be pleased to answer any questions during one of his tours through the cabin.

One hour after takeoff the DST drums sonorously westward in the last of the autumn twilight. Eight thousand feet below the land is enveloped in full night and only the electric fires of civilization maintain the reality of motion. Memphis is still six hours' flight time over the murky horizon.

Captain Dodson is an old-timer—helmet and goggles airmail. He wears two stripes on his sleeve, ostensibly of gold braid in the maritime style, but actually the material is a cheap imitation long since faded until it is only yellow ribbon. Dodson, who is balding, wears his uniform cap continuously when on duty. He clamps his headphones *over* his cap, enduring the entire awkward and heavy combination because like so many other veterans of open cockpit flying he feels undressed in an aircraft without something on his head. His copilot, still a namelsss and faceless nonperson as is the lot of all copilots, identified by one and one-half pseudogold stripes, keeps his left headphone away from that ear less his master offer priceless comments on the desirability of the stewardess and her possible willingness once they are returned to earth, or the effect of rumored aircraft additions on the copilot's seniority, or perhaps, some verbal jewel on the beauty of the night. And like all copilots sensitive to the fashion of their mentors he wears his cap throughout the flight.

"We have more crosswind than forecast. Figure me the wind."

The copilot checks his flight log for past position times and headings. He takes a round computer from his shirt pocket and turns it to the reverse side. Holding his small flashlight in his teeth he makes a pencil dot on the drift scale and twists the clear plastic disk until the dot lies across the true course line.

"I get ten degrees at twenty-five knots."

"Ah? We have a little help . . . maybe."

But doubting, Dodson looks down at the abyss below and meditates on the lights of a town sliding slowly beneath the left engine.

"No. I think we are in the middle of things. The wind is right on our beam."

And since Dodson is the captain that is where the wind is.

Later both Dodson and his copilot will be obliged to depart separately from the cockpit and make their way along the darkened aisle of the passenger cabin. It is common knowledge among airline personnel that regardless of uniform you can always differentiate an American Airlines pilot from those who fly for United. United equips their aircraft with Pratt and Whitney engines which are marvelously smooth in operation. Through corporate expediency American relies on Wright engines, faithful enough but cursed with such inherent vibrations the kidneys of all American pilots are sorely tried. Their frequent trips "to talk with the passengers" invariably terminate in the lavatory.

In spite of their unromantic mission aft "Mercury" pilots experience a curious emotion once they close the cockpit door. They may walk with a trifle less assurance, allow even a momentary hesitation in their silent parade. For there is no more stunning reminder of a man's aerial responsibility than to pass along the line of berths, the green curtains closed and swaying ever so slightly in response to the motion of the aircraft. Behind those curtains are individuals, trusting and utterly dependent on two total strangers who for this little time spell must directly control their destiny. The pilots are only ordinary men who must now relieve themselves.

If such brooding might create an atmosphere of melancholy in the cockpit, Dodson and his copilot can easily dismiss it. They may take refuge in minimizing their roles in the fast-developing scheme of the flying business.

"Full load again? I'll tell my relatives the company must be giving free rides to the coast."

"The Pullman cars must be hurting, but I guess it'll be ten years before I make captain."

The copilot's guess is deliberately pessimistic. If he keeps out of trouble there is every chance he will be wearing two full stripes after less than two years with the line.

As American Airlines pilots, both Dodson and his copilot fly for the "Flagship Fleet,"

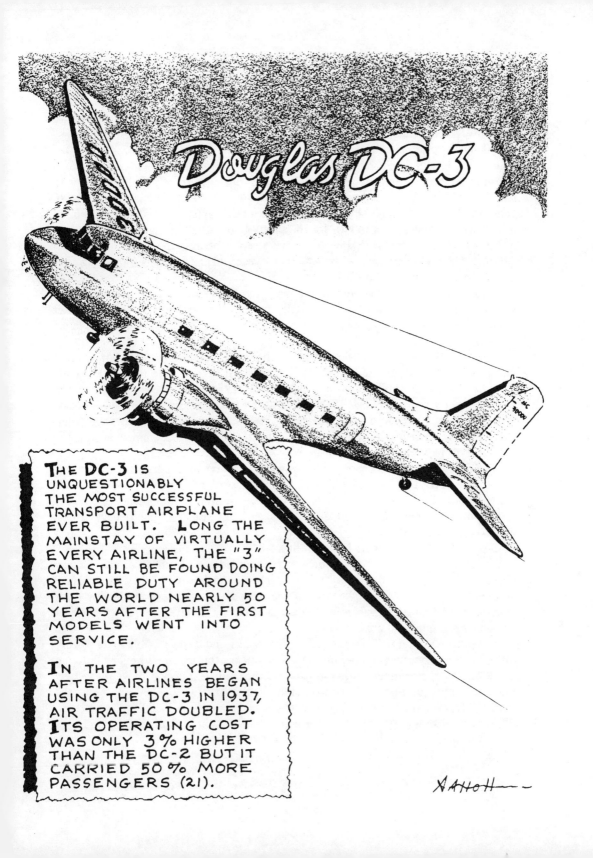

Douglas DC-3

THE DC-3 IS
UNQUESTIONABLY
THE MOST SUCCESSFUL
TRANSPORT AIRPLANE
EVER BUILT. LONG THE
MAINSTAY OF VIRTUALLY
EVERY AIRLINE, THE "3"
CAN STILL BE FOUND DOING
RELIABLE DUTY AROUND
THE WORLD NEARLY 50
YEARS AFTER THE FIRST
MODELS WENT INTO
SERVICE.

IN THE TWO YEARS
AFTER AIRLINES BEGAN
USING THE DC-3 IN 1937,
AIR TRAFFIC DOUBLED.
ITS OPERATING COST
WAS ONLY 3% HIGHER
THAN THE DC-2 BUT IT
CARRIED 50% MORE
PASSENGERS (21).

which affects certain trappings designed to impress customers and please news photographers. Each DST is named after a state of the union, and a flag bearing the double eagle insignia of American is hoisted above the right cockpit window. The "colors" ceremony is carried out by the copilot who opens his window immediately after the landing roll is completed and inserts the flagstaff in a special socket. Huzzah! But copilots are very human and subject to many distractions even if they are dignified with the title of "First Officer." And sometimes they forget to douse colors before takeoff, creating an unmistakable coolness in the attitude of the captain. In his adamant opinion a man who is paid $190 a month for the privilege of sitting on his right should not forget — *anything*. Captain Dodson, who flies sleepers because the pay is higher than daytime schedules, earns as much as $9,000 a year and if *he* forgets something only God, his wife, or his chief pilot dare accuse him.

It is a clear night and Dodson will follow the long line of winking airway lights into Memphis. Having made a wheel landing in the new style rather than a three-pointer which might awaken his passengers, he will sigh and sign the logbook with his name and seniority number.

Now, yawning mightily, Dodson pulls his flight bag from the baggage compartment just aft the copilot's seat and proceeds to the cabin. If he forgets to remove his name plate from the cabin bulkhead, the copilot or the stewardess will do it for him and moving quietly down the aisle they will descend into the soft Tennessee night one after the other. They are met for briefing by the new crew who will take the "Mercury" through to Dallas. And even outside the aircraft their voices remain subdued.

"The gyro precesses. I wrote it up."

"The man in number six snores like a walrus. The woman in upper three says wake her so she can see the dawn. Number eight is a please-do-not-disturb."

Memphis to Dallas and thence to Phoenix where the woman in upper three becomes ecstatic over the newborn desert day. And once aloft again yet another stewardess shakes each green curtain decorously.

"Breakfast will be served in forty minutes — "

Two hours later, but twenty minutes over scheduled arrival time, fourteen reasonably contented passengers disembark at Los Angeles-Glendale airport. They are refreshed by their long night's sleep and ready for the new day — which is considerably more than can be said for the irritable time-lagged passengers who have endured a flight over the same geography in a present-day jet.

The Douglas sleeper (DST) was comfort with wings, and it was possible to treat every passenger as a very important person. As a consequence the appeal of the long Pullman journey across the continent began to fade and thus did air transport move into a new and vitally important phase. Very soon after the inauguration of the DST flights the sister DC-3s were in the air plying their trade so efficiently no established line could afford to be without at least a few. Eight hundred and three commercial-type DC-3s were eventually built, and over 10,000 military versions.

The DSTs (née DC-3s) are an easy aircraft to fly, almost totally forgiving to the most ham-handed pilots. Their inherent stability makes them an excellent instrument aircraft and their low stall speed combined with practically full control response at slow approach speeds allows the use of very short fields. They can be slipped with full flaps or held nose high and allowed to descend in a near power stall. In the hands of a skilled pilot DC-3s can be successfully landed in just about any cabbage patch some optimist has dared to call an airfield.

Actual DC-3 performance records have caused frequent head shaking among those who could hardly believe what they knew was fact. Civil type DC-3s were licensed to gross 25,500 pounds. When World War II came along both ground and flying people became ever more casual with weights stuffed in airplanes. A gross of 31,500 pounds was not at all uncommon for DC-3s (C-47 in the most prolific military version), and no one bothered with such trimmings as weight and balance forms. Except for the placement of spare engines, which were a highly concentrated weight per cubic foot, the rule was to "shove everything as far forward as it will go." Once a China National DC-3 designed like all others to carry twenty-one passengers carried *seventy-five* passengers out of China to a Burmese air field. Among them was a certain James H. Doolittle who had commenced his journey on the aircraft carrier *Hornet* and was thence routed via Tokyo to an unscheduled landing on the Chinese mainland.

More than one thousand DC-3s (C-47s) took to the air on D-Day in 1944 and they were among those present at all the major World War II engagements. They kept right on flying throughout the Berlin airlift and continued through the Korean conflict.

DC-3s have been put to almost every conceivable aviation use. They have been used as makeshift bombers, glider tows, forest-fire bombers, ambulance litter ships operating directly out of battle zones, and equipped with rapid-fire guns they have even served as fighters in Vietnam. Most airline pilots referred to them as "Threes," the Air Force dubbed them "Gooney Birds," and the Navy, "R-4Ds." The British chose "Dakotas," and the Russians, who were presented with a great number, conveniently forgot where they were built and were soon manufacturing their own version.

Then there was the famous "DC-2½," a crippled CNAC aircraft, which flew to its home base quite sedately with a DC-3 wing on one side and a DC-2 wing on the other. There were more than fifty different military versions descended from the original DST.

Regardless of the label the Douglas masterpiece was profoundly admired by all who flew it.

And still is. Wherever you go in the world there is a very good chance you will see a DC-3 at work, and in some of the less sophisticated regions there is a very strong possibility you will be in one.

When John H. Glenn, Jr., became the first American to orbit the earth on February 20, 1962, the United States was still trying to narrow the lead which the Soviet Union had taken with its launching of *Sputnik* in 1957. Some ten months before Glenn's flight, Yuri Gagarin had become the first man in orbit; four months later, Gherman Titov had registered seventeen orbits. As 135 million Americans watched on television or listened on radio, Glenn's *Friendship* 7 capsule circled the globe three times. We were still behind, but closing in fast. This account appeared in *Life* twelve days after the flight.

JOHN H. GLENN, JR.

Earth Orbit

The two big outboard booster engines shut down and dropped away. We were out of the atmosphere by now, and had built up enough speed so that all we needed was the long, final push from the sustainer engine to drive us into orbit. There was no sensation of speed, however, because there was nothing outside to look at as a reference point. . . .

All during this period the Atlas was turning a corner in the sky. It was programing over in such a way that at one point the capsule was actually riding lower than the tail of the booster. I caught a quick glimpse of the Atlantic through the window. We were still climbing and I knew that we would pitch up again as we approached the fourth hurdle, the insertion into orbit.

The insertion was perfect. The sustainer engine cut off. Then, right on time, the bolts exploded off the ring that held the booster and the capsule together and the posigrade rockets fired to push the capsule out ahead by itself. The periscope extended and the capsule began its automatic turn—around to the position it would ride through the orbits, blunt end facing the direction we traveled. Now I could look back down out the window the way I'd come.

The view was tremendous. I could see for hundreds of miles in every direction—the sun on white clouds, patches of blue water beneath and great chunks of Florida and the southeastern United States. Much nearer I could see the booster drifting along by itself, not much more than 100 yards behind me and slightly above. But these were only quick glances. I had to concentrate on monitoring the capsule system. I could feel the zero G sensation now as the capsule and I flew weightless. Then Al Shepard called with the message I'd been waiting for. "You are go," he said, "for at least seven orbits." I was really jubilant. Shepard meant that the Cape computers indicated the insertion of the capsule was good enough for a minimum of seven orbits. It probably would have been good enough for seventeen or seventy if we had been able to carry enough fuel and oxygen for such a mission.

I loosened my chest strap now and went to work. There was much to be done. The plan was for me to spend most of the first orbit getting used to the new environment and helping the ground stations establish the best pattern of communications and radar tracking so we would be ready for any emergencies. I gave the controls—both manual and automatic—a thorough checkout on the way across the Atlantic and they responded perfectly. I was very happy to see this, for there is always some doubt whether the pilot and such complicated controls will work as well under actual conditions in space as they do on the trainers. I could see no difference, at least not yet.

I also started to make a few observations out the window. I spotted the Canary Islands about fifteen minutes after liftoff and picked up the African coast a couple of minutes later. The Atlas Mountains were clear through the window. Back inland I could see huge dust storms blowing across the desert as well as great clouds of smoke from brush fires along the edge of the desert. I reported the dust storms and the tracking station in Kano, Nigeria, told me they had been going on for a week. The first sunset, which came over the Indian Ocean, was a beautiful display of vivid colors—oranges, yellows, purples of all gradations—that extended out through the atmosphere for about 60 degrees on either side of the sun. The sun itself was so bright that I had to use filters to look directly at it. When the sun was still high in the sky its light was more bluish-white than yellow. When it came through the window, it was similar in color and intensity to the huge arc lights we use at the Cape.

I could see a few stars even by day, against the black sky. I observed them best at night and I was roughly able to determine my position with reference to familiar groups such as Orion or the Pleiades. Each time around I noticed a strange phenomenon. The stars shone steadily as they neared the horizon. Then they dimmed for a bit. But the stars brightened again before actually setting. They appeared to be passing through a layer of haze about six to eight degrees above the earth and two degrees thick. Such a haze layer is new to our scientists and we will be trying for a better look on the next flights. Just like the sun, incidentally, the stars set 18 times faster from my fast-moving, orbital vantage point than they do for us here on earth.

One of the things that surprised me most about the flight was the percentage of the earth covered by clouds. The clouds were nearly solid over central Africa and extended over most of the Indian Ocean. Western Australia was clear but it was cloudy again from eastern Australia almost all the way across the Pacific. I could not establish the exact altitude of the clouds, but I could tell the difference between the different layers from 150 miles up. The moon, which was almost full, was out each time I crossed the Pacific, and the clouds showed up crisp and clear in its light. Like the stars I saw, the moon looked little brighter than it would through the desert air on a clear night. But the window cut some of the light.

I saw a rare lighting effect as I crossed the Indian Ocean. There were large storms north of my course, and though I was nearly three quarters of a million feet above the clouds, I could clearly see lightning in them. The cloud interiors lit up as if they held bulbs which pulsed on and off. As I came over Australia on the first pass, I saw my first signs of manmade light. Out the window I could see several great patches of brightness down below where the citizens of Perth and several other cities on the western coast had turned on their lights to send me a greeting in electricity. It was a warm and welcome sight. I asked Astronaut Gordon Cooper, who was capsule communicator there, to express my thanks to the Australian people.

The strangest sight of all came with the very first ray of sunrise as I was crossing the Pacific toward the United States. I was checking the instrument panel and when I looked back out the window I thought for a minute that I must have tumbled upside-down and was looking up at a new field of stars. I checked my instruments to make sure I was right-side-up. Then I looked again. There, spread out as far as I could see, were literally thousands of tiny luminous objects that glowed in the black sky like fireflies. I was riding slowly through them, and the sensation was like walking backwards through a pasture where someone had waved a wand and made all the fireflies stop right where they were and glow steadily. They were greenish yellow in color, and they appeared to be about six to ten feet apart. I seemed to be passing through them at a speed of from three to five miles an hour. They were all around me, and those nearest the capsule would occasionally move across the window as if I had slightly interrupted their flow. On the next pass I turned the capsule around so that I was looking right into the flow, and though I could see far fewer of them in the light of the rising sun, they were still there. Watching them come toward me, I felt certain they were not caused by anything emanating from the capsule. I thought perhaps I'd stumbled into the lost batch of needles the air force had tried to set up in orbit for communications purposes. But I could think of no reason why needles should glow like fireflies, nor did they look like needles. As far as I know, the true identity of these particles is still a mystery.

James Norman Hall was an American who flew with the Lafayette Escadrille. In *High Adventure* (1918) he described his great escape from seven German fighters: shot through the lungs, he fell for over two miles, his motor going full speed all the way; then he miraculously regained consciousness just long enough to level out at 300 feet and glide in to safety. In 1920 he collaborated with Charles Nordhoff to write *The Lafayette Flying Corps* and in 1927 to write *Falcons of France*, from which the passage below is excerpted. At the end of the World War I, Hall had moved to Tahiti, which probably inspired his notable collaborations with Nordhoff in the 1930s: *Mutiny on the Bounty*, *Men Against the Sea*, and *Pitcairn Island*.

JAMES NORMAN HALL
and CHARLES NORDHOFF

Baptism of Fire

While in training in the schools I had often tried to imagine what my first air battle would be like. I haven't a very fertile imagination, and in my mental picture of such a battle I had seen planes approaching one another more or less deliberately, their guns spitting fire, then turning to spit again. That, in fact, is what happens, except that the approach is anything but deliberate once the engagement starts. But where I had been chiefly mistaken was in thinking of them fighting at a considerable distance from each other—two, or three, or even five hundred yards. The reality was far different. At the instant when I found myself surrounded by planes, I heard unmistakably the crackle of machine-gun fire. It is curious how different this sounds in the air when one's ears are deafened by altitude, the rush of wind, and the roar of the motor. Even when quite close it is only a faint crackle, but very distinct, each explosion impinging sharply on the eardrums. I turned my head over my shoulder, to breathe the acrid smoke of tracer bullets, and just then—whang! crash!—my windshield was shattered. I made a steep bank in time to see the black crosses of a silver-bellied Albatros turned up horizontally about twenty yards distant, as though the German pilot merely wanted to display them to convince me that he was really a German. Then, as I leveled off, glancing hastily to my right, I saw not 10 meters below my altitude and flying in the same direction a craft that looked enormous, larger than three of mine. She had staggered wings, and there was no doubt about the insignia on her fishlike tail: that, too, was a black cross. It was a two-seater, and so close that I could clearly see the pilot and the gunner in the back seat. Body and wings were camouflaged, not in daubs after the French fashion, but in zigzag lines of brown and green. The observer, whose back was toward me, was aiming two guns mounted on a single swivel on the circular tract surrounding his cockpit. He crouched down, firing at a steep angle at someone overhead whom I could not see, his tracers stabbing through the air in thin clear lines. Apparently neither the pilot nor the rear gunner saw me. Then I had a blurted glimpse of the tricolor *cocardes* of a Spad that passed me like a flash, going in the opposite direction; and in that same instant I saw another Spad appear directly under the two-seater, nose up vertically, and seem to hang there as though suspended by an invisible wire.

What then happened is beyond the power of any words of mine to describe. A sheet of intense flame shot up from the two-seater, lapping like water around the wings and blown back along the body of the plane. The observer dropped his guns and I could all but see the expression of horror on his face as he turned. He ducked for a second with his arms around his head in an effort to protect himself; then without a moment's hesitation he climbed on his seat and threw himself off into space. The huge plane veered up on one side, turned nose down, and disappeared beneath me. Five seconds later I was alone. There wasn't another plane to be seen.

Catch-22. The phrase has now passed into the language, but it started with the scene in Doc Daneeka's tent reprinted below. And why was Yossarian so desperate to get out of combat duty? The "milk run"—that term so fraught with peril for any flight crew—over Bologna gives the answer.

JOSEPH HELLER

Crazy Over Bologna

"You're wasting your time," Doc Daneeka was forced to tell him.

"Can't you ground someone who's crazy?"

"Oh, sure. I have to. There's a rule saying I have to ground anyone who's crazy."

"Then why don't you ground me? I'm crazy. Ask Clevinger."

"Clevinger? Where *is* Clevinger? You find Clevinger and I'll ask him."

"Then ask any of the others. They'll tell you how crazy I am."

"They're crazy."

"Then why don't you gound them?"

"Why don't they ask me to ground them?"

"Because they're crazy, that's why."

"Of course they're crazy," Doc Daneeka replied. "I just told you they're crazy, didn't I? And you can't let crazy people decide whether you're crazy or not, can you?"

Yossarian looked at him soberly and tried another approach. "Is Orr crazy?"

"He sure is," Doc Daneeka said.

"Can you ground him?"

"I sure can. But first he has to ask me to. That's part of the rule."

"Then why doesn't he ask you to?"

"Because he's crazy," Doc Daneeka said. "He has to be crazy to keep flying combat missions after all the close calls he's had. Sure, I can ground Orr. But first he has to ask me to."

"That's all he has to do to be grounded?"

"That's all. Let him ask me."

"And then you can ground him?" Yossarian asked.

"No. Then I can't ground him."

"You mean there's a catch?"

"Sure there's a catch," Doc Daneeka replied. "Catch-22. Anyone who wants to get out of combat duty isn't really crazy."

There was only one catch and that was Catch-22, which specified that a concern for one's own safety in the face of dangers that were real and immediate was the process of a rational mind. Orr was crazy and could be grounded. All he had to do was ask; and as soon as he did, he would no longer be crazy and would have to fly more missions. Orr would be crazy to fly more missions and sane if he didn't, but if he was sane he had to fly them. If he flew them he was crazy and didn't have to; but if he didn't want to he was sane and had to. Yossarian was moved very deeply by the absolute simplicity of this clause of Catch-22 and let out a respectful whistle.

"That's some catch, that Catch-22," he observed.

"It's the best there is," Doc Daneeka agreed. . . .

Captain Piltchard and Captain Wren, the inoffensive joint squadron operations officers, were both mild, soft-spoken men of less than middle height who enjoyed flying combat missions and begged nothing more of life and Colonel Cathcart than the opportunity to continue flying them. They had flown hundreds of combat missions and wanted to fly hundreds more. They assigned themselves to every one. Nothing so wonderful as war had ever happened to them before; and they were afraid it might never happen to them again. They conducted their duties humbly and reticently, with a minimum of fuss, and went to great lengths not to antagonize anyone. They smiled quickly at everyone they passed. When they spoke, they mumbled. They were shifty, cheerful, subservient men who were comfortable only with each other and never met anyone else's eye, not even Yossarian's eye at the open-air meeting they called to reprimand him publicly for making Kid Sampson turn back from the mission to Bologna.

"Fellas," said Captain Piltchard, who had thinning dark hair and smiled awkwardly. "When you turn back from a mission, try to make sure it's for something important, will you? Not for something unimportant . . . like a defective intercom . . . or something like that. Okay? Captain Wren has more he wants to say to you on that subject."

"Captain Piltchard's right, fellas," said Captain Wren. "And that's all I'm going to say to you on that subject. Well, we finally got to Bologna today, and we found out it's a milk run. We were all a little nervous, I guess, and didn't do too much damage. Well, listen to this. Colonel Cathcart got permission for us to go back. And tomorrow we're really going to paste those ammunition dumps. Now, what to you think about that?"

And to prove to Yossarian that they bore him no animosity, they even assigned him to fly lead bombardier with McWatt in the first formation when they went back to Bologna the next day. He came in on the target like a Havermeyer confidently taking no evasive action at all, and suddenly they were shooting the living shit out of him!

Heavy flak was everywhere! He had been lulled, lured, and trapped, and there was nothing he could do but sit there like an idiot and watch the ugly black puffs smashing up to kill him. There was nothing he could do until his bombs dropped but look back into the bombsight, where the fine cross-hairs in the lens were glued magnetically over the target exactly where he had placed them, intersecting perfectly deep inside the yard of his block of camouflaged warehouses before the base of the first building. He was trembling steadily as the plane crept ahead. He could hear the hollow *boom-boom-boom-boom* of the flak pounding all around him in overlapping measures of four, the sharp, piercing *crack!* of a single shell exploding suddenly very close by. His head was busting with a thousand dissonant impulses as he prayed for the bombs to drop. He wanted to sob. The engines droned on monotonously like a fat, lazy fly. At last the indices on the bombsight crossed, tripping away the eight 500-pounders one after the other. The plane lurched upward buoyantly with the lightened load. Yossarian bent away from the bombsight crookedly to watch the indicator on his left. When the pointer touched zero, he closed the bomb bay doors and, over the intercom, at the very top of his voice, shrieked:

"Turn right hard!"

McWatt responded instantly. With a grinding howl of engines, he flipped the plane over on one wing and wrung it around remorselessly in a screaming turn away from the twin spires of flak Yossarian had spied stabbing toward them. Then Yossarian had McWatt climb and keep climbing higher and higher until they tore free finally into a calm, diamond-blue sky that was sunny and pure everywhere and laced in the distance with long white veils of tenuous fluff. The wind strummed soothingly against the cylindrical panes of his windows, and he relaxed exultantly only until they picked up speed again and then turned McWatt left and plunged him right back down, noticing with a transitory spasm of elation the mushrooming clusters of flak leaping open high above him and back over his shoulder to the right, exactly where he could have been if he had not turned left and dived. He leveled McWatt out with another harsh cry and whipped him upward and around again into a ragged blue patch of unpolluted air just as the bombs he had dropped began to strike. The first one fell in the yard, exactly where he had aimed, and then the rest of the bombs from his own plane and from the other planes in his flight burst open on the ground in a charge of rapid orange flashes across the tops of the buildings, which collapsed instantly in a vast, churning wave of pink and gray and coal-black smoke that went rolling out turbulently in all directions and quaked convulsively in its bowels as though from great blasts of red and white and golden sheet lightning.

"Well, will you look at that," Aarfy marveled sonorously right beside Yossarian, his plump, orbicular face sparkling with a look of bright enchantment. "There must have been an ammunition dump down there."

Yossarian had forgotten about Aarfy. "Get out!" he shouted at him. "Get out of the nose!"

Aarfy smiled politely and pointed down toward the target in a generous invitation for Yossarian to look. Yossarian began slapping at him insistently and signaled wildly toward the entrance of the crawlway.

"Get back in the ship!" he cried frantically. "Get back in the ship!"

Aarfy shrugged amiably. "I can't hear you," he explained.

Yossarian seized him by the straps of his parachute harness and pushed him backward toward the crawlway just as the plane was hit with a jarring concussion that rattled his bones and made his heart stop. He knew at once they were all dead.

"Climb!" he screamed into the intercom at McWatt when he saw he was still alive. *"Climb, you bastard! Climb, climb, climb, climb!"*

The plane zoomed upward again in a climb that was swift and straining, until he leveled it out with another harsh shout at McWatt and wrenched it around once more in a roaring, merciless forty-five-degree turn that sucked his insides out in one enervating sniff and left him floating fleshless in mid-air until he leveled McWatt out again just long enough to hurl him back around toward the right and then down into a screeching dive. Through endless blobs of ghostly black smoke he sped, the hanging smut wafting against the smooth plexiglass nose of the ship like an evil, damp, sooty vapor against his cheeks. His heart was hammering again in aching terror as he hurtled upward and downward through the blind gangs of flak charging murderously into the sky at him, then sagging inertly. Sweat gushed from his neck in torrents and poured down over his chest and waist with the feeling of warm slime. He was vaguely aware for an instant that the planes in his formation were no longer there, and then he was aware of only himself. His throat hurt like a raw slash from the strangling intensity with which he shrieked each command to McWatt. The engines rose to a deafening, agonized, ululating bellow each time McWatt changed direction. And far out in front the bursts of flak were still swarming into the sky from new batteries of guns poking around for accurate altitude as they waited sadistically for him to fly into range.

The plane was slammed again suddenly with another loud, jarring explosion that almost rocked it over on its back, and the nose filled immediately with sweet clouds of blue smoke. *Something was on fire!* Yossarian whirled to escape and smacked into Aarfy, who had struck a match and was placidly lighting his pipe. Yossarian gaped at his grinning, moon-faced navigator in utter shock and confusion. It occurred to him that one of them was mad.

"Jesus Christ!" he screamed at Aarfy in tortured amazament. "Get the hell out of the nose! Are you crazy? Get out!"

"What?" said Aarfy.

"Get out!" Yossarian yelled hysterically, and began clubbing Aarfy backhanded with both fists to drive him away. "Get out!"

"I still can't hear you," Aarfy called back innocently with an expression of mild and reproving perplexity. "You'll have to talk a little louder."

"Get out of the nose!" Yossarian shrieked in frustration. "They're trying to kill us! Don't you understand? They're trying to kill us!"

"Which way should I go, goddamn it?" McWatt shouted furiously over the intercom in a suffering, high-pitched voice. "Which way should I go?"

"Turn left! *Left*, you goddam dirty son of a bitch! Turn left *hard*!"

Aarfy crept up close behind Yossarian and jabbed him sharply in the ribs with the stem of his pipe. Yossarian flew up toward the ceiling with a whinnying cry, then jumped completely around on his knees, white as a sheet and quivering with rage. Aarfy winked encouragingly and jerked his thumb back toward McWatt with a humorous *moue*.

"What's eating *him*?" he asked with a laugh.

Yossarian was struck with a weird sense of distortion. "Will you get out of here?" he yelped beseechingly, and shoved Aarfy over with all his strength. "Are you deaf or something? Get back in the plane!" And to McWatt he screamed, "Dive! *Dive*!"

Down they sank once more into the crunching, thudding, voluminous barrage of bursting antiaircraft shells as Aarfy came creeping back behind Yossarian and jabbed him sharply in the ribs again. Yossarian shied upward with another whinnying gasp.

"I still couldn't hear you," Aarfy said.

"I said get *out of here*!" Yossarian shouted, and broke into tears. He began punching Aarfy in the body with both hands as hard as he could. "Get *away* from me! Get *away*!"

Punching Aarfy was like sinking his fists into a limp sack of inflated rubber. There was no resistance, no response at all from the soft, insensitive mass, and after a while Yossarian's spirit died and his arms dropped helplessly with exhaustion. He was overcome with a humilitating feeling of impotence and was ready to weep in self-pity.

"What did you say?" Aarfy asked.

"Get *away* from me," Yossarian answered, pleading with him now, "Go back in the plane."

"I still can't hear you."

"Never mind," wailed Yossarian, "never mind. Just leave me alone."

"Never mind what?"

Yossarian began hitting himself in the forehead. He seized Aarfy by the shirt front and, struggling to his feet for traction, dragged him to the rear of the nose compartment and flung him down like a bloated and unwieldy bag in the entrance of the crawlway. A shell banged open with a stupendous clout right beside his ear as he was scrambling back toward the front, and some undestroyed recess of his intelligence wondered that it did not kill them all. They were climbing again. The engines were howling again as though in pain, and the air inside the plane was acrid with the smell of machinery and fetid with the stench of gasoline. The next thing he knew, *it was snowing*!

Thousands of tiny bits of white paper were falling like snowflakes inside the plane, milling around his head so thickly that they clung to his eyelashes when he blinked in astonishment and fluttered against his nostrils and lips each time he inhaled. When he spun around in bewilderment, Aarfy was grinning proudly from ear to ear like something inhuman as he held up a shattered paper map for Yossarian to see. A large chunk of flak had ripped up from the floor through Aarfy's colossal jumble of maps and had ripped out through the ceiling inches away from their heads. Aarfy's joy was sublime.

"Will you look at this?" he murmured, waggling two of his stubby fingers playfully into Yossarian's face through the hole in one of his maps. "Will you look at this?"

Yossarian was dumfounded by his state of rapturous contentment. Aarfy was like an eerie ogre in a dream, incapable of being bruised or evaded, and Yossarian dreaded him for a complex of reasons he was too petrified to untangle. Wind whistling up through the jagged gash in the floor kept the myriad bits of paper circulating like alabaster particles in a paperweight and contributed to a sensation of lacquered, waterlogged unreality. Everything seemed strange, so tawdry and grotesque. His head was throbbing from a shrill clamor that drilled relentlessly into both ears. It was McWatt, begging for directions in an incoherent frenzy. Yossarian continued staring in tormented fascination at Aarfy's spherical countenance beaming at him so serenely and vacantly through the drifting whorls of white paper bits and concluded that he was a raving lunatic just as eight bursts of flak broke open successively at eye level off to the right, then eight more, and then eight more, the last group pulled over toward the left so that they were almost directly in front.

"Turn left hard!" he hollered to McWatt, as Aarfy kept grinning, and McWatt did turn left hard, but the flak turned left hard with them, catching up fast, and Yossarian hollered, "I said *hard, hard, hard, hard, you bastard, hard!*"

And McWatt bent the plane around even harder still, and suddenly, miraculously, they were out of range. The flak ended. The guns stopped booming at them. And they were alive.

Behind him, men were dying. Strung out for miles in a stricken, tortuous, squirming line, the other flights of planes were making the same hazardous journey over the target, threading their swift way through the swollen masses of new and old bursts of flak like rats racing in a pack through their own droppings. One was on fire, and flapped lamely off by itself, billowing gigantically like a monstrous blood-red star. As Yossarian watched, the burning plane floated over on its side and began spiraling down slowly in wide, tremulous, narrowing circles, its huge flaming burden blazing orange and flaring out in back like a long, swirling cape of fire and smoke. There were parachutes, one, two, three . . . four, and then the plane gyrated into a spin and fell the rest of the way to the ground, fluttering insensibly inside its vivid pyre like a shred of colored tissue paper. One whole flight of planes from another squadron had been blasted apart.

Yossarian sighed barrenly, his day's work done. He was listless and sticky. The engines crooned mellifluously as McWatt throttled back to loiter and allow the rest of the planes in his flight to catch up. The abrupt stillness seemed alien and artificial, a little insidious. Yossarian unsnapped his flak suit and took off his helmet. He sighed again, restlessly, and closed his eyes and tried to relax.

"Where's Orr?" someone asked suddenly over his intercom.

Yossarian bounded up with a one-syllable cry that crackled with anxiety and provided the only rational explanation for the whole mysterious phenomenon of the flak at Bologna: Orr! He lunged forward over the bombsight to search downward through the plexiglass for some reassuring sign of Orr, who drew flak like a magnet and who had undoubtedly attracted the crack batteries of the whole Hermann Goering Division to Bologna overnight from wherever the hell they had been stationed the day before when Orr was still in Rome. Aarfy launched himself forward an instant later and cracked Yossarian on the bridge of the nose with the sharp rim of his flak helmet. Yossarian cursed him as his eyes flooded with tears.

"There he is," Aarfy orated funereally, pointing down dramatically at a hay wagon and two horses standing before the barn of a gray stone farmhouse. "Smashed to bits. I guess their numbers were all up."

Yossarian swore at Aarfy again and continued searching intently, cold with a compassionate kind of fear now for the little bouncy and bizarre buck-toothed tentmate who had smashed Appleby's forehead open with a ping-pong racket and who was scaring the daylights out of Yossarian once again. At last Yossarian spotted the two-engined, twin-ruddered plane as it flew out of the green background of the forests over a field of yellow farmland. One of the propellers was feathered and perfectly still, but the plane was maintaining altitude and holding a proper course. Yossarian muttered an unconscious prayer of thankfulness and then flared up at Orr savagely in a ranting fusion of resentment and relief.

"That bastard!" he began. "That goddam stunted, red-faced, big-cheeked, curly-headed, buck-toothed rat bastard son of a bitch!"

"What?" said Aarfy.

"That dirty goddam midget-assed, apple-cheeked, goggle-eyed, undersized, buck-toothed, grinning, crazy sonofabitchinbastard!" Yossarian sputtered.

"What?"

"Never mind!"

"I still can't hear you," Aarfy answered.

Yossarian swung himself around methodically to face Aarfy. "You prick," he began.

"Me?"

"You pompous, rotund, neighborly, vacuous, complacent . . ."

Aarfy was urperturbed. Calmly he struck a wooden match and sucked noisily at his pipe with an eloquent air of benign and magnanimous forgiveness. He smiled sociably and opened his mouth to speak. Yossarian put his hand over Aarfy's mouth and pushed him away wearily. He shut his eyes and pretended to sleep all the way back to the field so that he would not have to listen to Aarfy or see him.

At the briefing room Yossarian made his intelligence report to Captain Black and then waited in muttering suspense with all the others until Orr chugged into sight overhead finally with his one good engine still keeping him aloft gamely. Nobody breathed. Orr's landing gear would not come down. Yossarian hung around only until Orr had crash-landed safely, and then stole the first jeep he could find with a key in the ignition and raced back to his tent to begin packing feverishly for the emergency rest leave he had decided to take in Rome, where he found Luciana and her invisible scar that same night.

On October 25, 1944, the U.S. escort carrier *St. Lo* was struck by a diving Japanese plane whose pilot had crashed into it with full intent. This was the first kamikaze attack. In the ensuing months 1,228 Japanese pilots flew such suicide missions, sinking 34 ships and damaging 288 more, killing many. Were these pilots madmen, hypnotized by heartless commanders? No; they were brave, honorable, remarkably dispassionate men whose devotion to their country and their emperor knew no bounds. The letters which follow are from *The Divine Wind*, which is the translation of the word "kamikaze." In 1281, the Mongol emperor Kublai Khan prepared an armada to invade Japan, but his fleet was destroyed by a great typhoon off the Japanese coast; the Japanese regarded this timely storm as heavenly intervention — a divine wind. When their nation seemed on the brink of defeat in late 1944, they looked to the divine wind once more.

RIKIHEI INOGUCHI, TADASHI NAKAJIMA, and ROGER PINEAU

Blossoms in the Wind

> *In blossom today, then scattered:*
> *Life is so like a delicate flower.*
> *How can one expect the fragrance*
> *To last forever?*
>
> —VICE ADMIRAL OHNISHI,
> KAMIKAZE SPECIAL ATTACK FORCE

What, then, were the thoughts and feelings of the suicide pilots themselves as they volunteered, waited their turn, and went out on their missions?

Mr. Ichiro Ohmi made a nationwide pilgrimage for four and a half years after the war to visit the homes of kamikaze pilots. The families showed him mementoes and letters of their loved ones. He has kindly provided [us] with copies of these letters, some of which express more clearly than could any other words the thoughts and feelings of the pilots about to die.

In general, what little the enlisted pilots wrote was of a simple, straightforward nature. Academy graduates also wrote very little — perhaps because they were thoroughly indoctrinated in the way of the warrior and thus accepted their fate matter-of-factly. It was the reserve officers from civilian colleges and universities, who had had only a hasty military training before receiving their assignments, who wrote the most.* A few typical letters serve to convey the spirit of kamikaze pilots.

The following letter is by Flying Petty Officer First Class Isao Matsuo of the 701st Air Group. It was written just before he sortied for a kamikaze attack. His home was in Nagasaki Prefecture.

OCTOBER 28, 1944

Dear Parents:

Please congratulate me. I have been given a splendid opportunity to die. This is my last day. The destiny of our homeland hinges on the decisive battle in the seas to the south where I shall fall like a blossom from a radiant cherry tree.

I shall be a shield for His Majesty and die cleanly along with my squadron leader and other friends. I wish that I could be born seven times, each time to smite the enemy.

How I appreciate this chance to die like a man! I am grateful from the depths of my heart to the parents who have reared me with their constant prayers and tender love. And I am grateful as well to my squadron leader and superior officers who have looked after me as if I were their own son and given me such careful training.

Thank you, my parents, for the twenty-three years during which you have cared for me and inspired me. I hope that my present deed will in some small way repay what you have done for me. Think well of me and know that your Isao died for our country. This is my last wish, and there is nothing else that I desire.

I shall return in spirit and look forward to your visit at the Yasukuni Shrine. Please take good care of yourselves.

How glorious is the Special Attack Corps' Giretsu Unit whose *Suisei* bombers will attack the enemy. Movie cameramen have been here to take our pictures. It is possible that you may see us in newsreels at the theater.

*It must be borne in mind that for many hundreds of years while the code of the warrior *(Bushido)*, which stressed as necessary a willingness to die at any moment, governed the conduct of the samurai, similar principles were concurrently adopted by merchants, farmers, and artisans, stressing the value of unquestioning loyalty to the emperor, other superiors, and the people of Japan. Thus, the introduction of the kamikaze principle was not so shocking to these Japanese as it would be to an Occidental. In addition, the belief that one continues to live, in close association with both the living and the dead, after death, generally causes their concept of death to be less final and unpleasant in its implications.

We are sixteen warriors manning the bombers. May our death be as sudden and clean as the shattering of crystal.

Written at Manila on the eve of our sortie.

Isao

Soaring into the sky of the southern seas, it is our glorious mission to die as the shields of His Majesty. Cherry blossoms glisten as they open and fall.

Lieutenant (jg) Nobuo Ishibashi, a native of Saga City in northern Kyushu, was born in 1920. He was a member of the Tsukuba Air Group before his assignment to the special attack corps. This is his last letter home.

Dear Father:

Spring seems to come early to southern Kyushu. Here the blossoms and flowers are all beautiful. There is a peace and tranquility, and yet this place is really a battleground.

I slept well last night; didn't even dream. Today my head is clear and I am in excellent health.

It makes me feel good to know that we are on the same island at this time.

Please remember me when you go to the temple, and give my regards to all of our friends.

Nobuo

I think of springtime in Japan while soaring to dash against the enemy.

The following letter was written by Ensign Ichizo Hayashi, born in 1922 in Fukuoka Prefecture of northern Kyushu. He had been reared in the Christian faith. Upon graduation from Imperial University at Kyoto he joined the Genzan (Wonsan) Air Group, from which he was assigned to the special attack corps.

Dearest Mother:

I trust that you are in good health.

I am a member of the *Shichisei* Unit of the special attack corps. Half of our unit flew to Okinawa today to dive against enemy ships. The rest of us will sortie in two or three days. It may be that our attack will be made on April 8, the birthday of Buddha.

We are relaxing in an officers' billet located in a former school building near the Kanoya air base. Because there is no electricity, we have built a roaring log fire and I am writing these words by its light.

Morale is high as we hear of the glorious successes achieved by our comrades who have gone before. In the evening I stroll through clover fields, recalling days of the past.

On our arrival here from the northern part of Korea we were surprised to find that cherry blossoms were falling. The warmth of this southern climate is soothing and comforting.

Please do not grieve for me, mother. It will be glorious to die in action. I am grateful to be able to die in a battle to determine the destiny of our country.

As we flew into Kyushu from Korea the route did not pass over our home, but as our planes approached the homeland I sang familiar songs and bade farewell to you. There remains nothing in particular that I wish to do or say, since Umeno will convey my last desires to you. This writing is only to tell you of the things that occur to me here.

Please dispose of my things as you wish after my death.

My correspondence has been neglected recently so I will appreciate it if you remember me to relatives and friends. I regret having to ask this of you, but there is now so little time for me to write.

Many of our boys are taking off today on their one-way mission against the enemy. I wish that you could be here in person to see the wonderful spirit and morale at this base.

Please burn all my personal papers, including my diaries. You may read them, of course, mother, if you wish, but they should not be read by other people. So please be sure to burn them after you have looked at them.

On our last sortie we will wear regular flight uniforms and a headband bearing the rising sun. Snow-white mufflers give a certain dash to our appearance.

I will also carry the rising sun flag which you gave to me. You will remember that it bears the poem, "Even though a thousand men fall to my right and ten thousand fall to my left. . . ." I will keep your picture in my bosom on the sortie, mother, and also the photo of Makio-san.

I am going to score a direct hit on an enemy ship without fail. When war results are announced you may be sure that one of the successes was scored by me. I am determined to keep calm and do a perfect job to the last, knowing that you will be watching over me and praying for my success. There will be no clouds of doubt or fear when I make the final plunge.

On our last sortie we will be given a package of bean curd and rice. It is reassuring to depart with such good luncheon fare. I think I'll also take along the charm and the dried bonito from Mr. Tateishi. The bonito will help me to rise from the ocean, mother, and swim back to you.

At our next meeting we shall have many things to talk about which are difficult to discuss in writing. But then we have lived together so congenially that many things may now be left unsaid. "I am living in a dream which will transport me from the earth tomorrow."

Yet with these thoughts I have the feeling that those who went on their missions yesterday are still alive. They could appear again at any moment.

The rocket-powered **Ohka Kamikaze** carried a 2,645-pound bomb, more than half of its total weight. Americans called it **BAKA** – Japanese for "fool."

Kamikaze

THE SPECIAL AIR FORCE WAS JAPAN'S LAST-DITCH ATTEMPT TO STOP AMERICAN ADVANCES BY USING VOLUNTARY SUICIDE AIR ATTACKS ON U.S. SHIPS.

DURING THE INVASION OF OKINAWA, THE JAPANESE USED MORE THAN 6,000 KAMIKAZES. ONLY A FEW HUNDRED ACTUALLY HIT SHIPS BUT THEY CAUSED GREAT DAMAGE AND MANY DEATHS.

In my case please accept my passing for once and for all. As it is said, "Let the dead past bury its dead." It is most important that families live for the living.

There was a movie shown recently in which I thought I saw Hakata. It gave me a great desire to see Hakata again just once before going on this last mission.

Mother, I do not want you to grieve over my death. I do not mind if you weep. Go ahead and weep. But please realize that my death is for the best, and do not feel bitter about it.

I have had a happy life, for many people have been good to me. I have often wondered why. It is a real solace to think that I may have some merits which make me worthy of these kindnesses. It would be difficult to die with the thought that one had not been anything in life.

From all reports it is clear that we have blunted the actions of the enemy. Victory will be with us. Our sortie will deliver a coup de grâce to the enemy. I am very happy.

We live in the spirit of Jesus Christ, and we die in that spirit. This thought stays with me. It is gratifying to live in this world, but living has a spirit of futility about it now. It is time to die. I do not seek reasons for dying. My only search is for an enemy target against which to dive.

You have been a wonderful mother to me. I only fear that I have not been worthy of the affection you have lavished on me. The circumstances of my life make me happy and proud. I seek to maintain the reason for this pride and joy until the last moment. If I were to be deprived of present surroundings and opportunities my life would be worth nothing. Standing alone, I was good for little. I am grateful, therefore, for the opportunity to serve as a man. If these thoughts sound peculiar, it is probably because I am getting sleepy. But for my drowsiness there are many other things I should like to say.

There is nothing more for me to say, however, by way of farewell.

I will precede you now, mother, in the approach to heaven. Please pray for my admittance. I should regret being barred from the heaven to which you will surely be admitted.

Pray for me, mother.

Farewell,

Ichizo

(When his sortie was delayed, this flier added the following postscript to his letter.)

"Strolling between the paddy fields the night is serene as I listen to the chant of the frogs." I could not help but think of this during my walk last evening. I lay down in a field of clover and thought of home. Upon my return to the barracks, my friends said that I smelled of clover and it brought them memories of home and mother. Several of them commented that I must have been a mamma's boy.

This did not disturb me at all; in fact, I was pleased by the remark. It is an index that people like me. When I am disturbed it is good to think of the many people who have been so kind to me, and I am pacified. My efforts will be doubled to prove my appreciation of the kindhearted people it has been my pleasure to know.

The cherry blossoms have already fallen. I wash my face each morning in a nearby stream. It reminds me of the blossom-filled stream that ran near our home.

It appears that we will go to make our attack tomorrow. Thus the anniversary of my death will be April 10. If you a service to commemorate me, I wish you to have a happy family dinner.

Now it is raining, the kind of rain we have in Japan rather than what I experienced in Korea. There is an old organ in our billet and someone is playing childhood songs, including the one about a mother coming to school with an umbrella for her child.

The departure was again postponed for this flier, and he had a chance to add yet another bit to the letter, which was finally mailed after he had taken off on his final flight:

I have thought that each day would be the last, but just as with most things in life, one can never be certain. It is the evening of April 11, and this was not my day.

Do hope that I was photogenic today, for several newsreel cameramen were here, and they singled me out for a special series of pictures. Later the Commander in Chief of Combined Fleet greeted us in our billet and said to me, "Please do your best." It was a great honor for me that he would speak to so humble a person as myself. He is convinced that the country's fate rests upon our shoulders.

Today we gathered about the organ and sang hymns.

Tomorrow I will plunge against the enemy without fail.

Ensign Heiichi Okabe was born in 1923. His home was Fukuoka Prefecture of northern Kyushu. Before enlisting he was graduated from Taihoku Imperial University. His first duty was in the Wonsan Air Group, and he was transferred thence to *Shichisei* Unit No. 2 of the special attack corps. He kept a diary which was sent to his family after his final sortie. The following is an excerpt from one of his last entries in that diary:

FEBRUARY 22, 1945

I am actually a member at last of the Kamikaze Special Attack Corps.

My life will be rounded out in the next thirty days. My chance will come! Death and I are waiting. The training and practice have been rigorous, but it is worthwhile if we can die beautifully and for a cause.

I shall die watching the pathetic struggle of our nation. My life will gallop in the next few weeks as my youth and life draw to a close. . . .

. . . The sortie has been scheduled for the next ten days.

I am a human being and hope to be neither saint nor scoundrel, hero nor fool—just a human being. As one who has spent his life in wistful longing and searching, I die resignedly in the hope that my life will serve as a "human document."

The world in which I live was too full of discord. As a community of rational human beings it should be better composed. Lacking a single great conductor, everyone lets loose with his own sound, creating dissonance where there should be melody and harmony.

We shall serve the nation gladly in its present painful struggle. We shall plunge into enemy ships cherishing the conviction that Japan has been and will be a place where only lovely homes, brave women, and beautiful friendships are allowed to exist.

What is the duty today? It is to fight.

What is the duty tomorrow? It is to win.

What is the daily duty? It is to die.

We die in battle without complaint. I wonder if others, like scientists, who pursue the war effort on their own fronts, would die as we do without complaint. Only then will the unity of Japan be such that she can have any prospect of winning the war.

If, by some strange chance, Japan should suddenly win this war, it would be a fatal misfortune for the future of the nation. It will be better for our nation and people if they are tempered through real ordeals which will serve to strengthen.

Like cherry blossoms
In the spring,
Let us fall
Clean and radiant.

And he lived to tell about it. Duke Krantz, more dramatically known as the Flying Diavolo, was a stunt flier for the Gates Flying Circus. Such circuses consisted of itinerant barnstormers who believed they could make more money by banding together than by winging it on their own, giving rides and exhibitions à la Lincoln Beachey. In that crazy era of flagpole sitters and human flies and all kinds of hair-raising escapades, nothing gave rise to more butterflies in the stomach than watching a wing walker.

DUKE KRANTZ

Those Daredevil Stunts of Yesteryear

We had crowds sometimes of 30,000 or 40,000 people—tremendous crowds. The main thing was, of course, to put on wing-walking shows to draw the people. Then, in between times, we would sell tickets and take the people for rides. Of course, it was very scientifically arranged so that we could load and unload rapidly. One man would pull passengers out from one side of the ship while another man would push them in from the opposite side.

In the beginning, we charged $2.50 for a short ride, $5 for a long ride. If they wanted any stunts at all, it would be an additional $5. There was a dollar admission fee for the exhibition, but the stunts were just the attraction to get passengers.

At first we used to sit on the top wing of the plane with our toes underneath, sort of around the straps. But, of course, that was not as spectacular as standing up on top. In order to do that we had to have little secret devices, like having a belt under my sweater that nobody could see. I had cables—full cables, not wires—made up with snaps on them, and they were hidden in the airplane all the time. So when I got up in the air, I would stand up, I would hook the cables to the belt, and take one in each hand. As I reached up, I was snapping these fittings that nobody could see. Then I would straighten up and, with no hesitation at all, stretch out good and tight. The pilot would go into a dive and come around in a loop. That was, of course, one of the spectacular stunts.

The breakaway was quite exciting too. We had a very narrow cable ladder so you could pull yourself up onto the wing. We'd put straps around the ankles and snap the snaps at the end of the cables onto them. At the right time we would dive from the landing gear, head down, swing for a while, then come back up and climb up the ladder.

For plane changes—of course, you had to use two airplanes—the ladder would be attached to the wings of the ship above. I would climb out on the wings of the lower ship and walk out past the outer strut and up on the upper wing, where there was a brace that you could steady your leg against. That was the only thing you could hold on to at all. The upper ship would overtake the lower one with the ladder swinging from its wing tip, and I would reach for the last rung. Then it was only a matter of hanging on, because by that time the other ship was out from under you, leaving you in space. It was quite exciting. One time the ships came too close, and to keep from being hit I threw myself down on the wings and lay with my shoulder over the aileron. The other wing from the upper plane was not more than two feet from my face. By that time they started to separate. I grabbed for the ladder and got hold of the center part of it, but it so happened that the lower end of the ladder got caught between the aileron and wing of the lower plane. There I was, hanging in the middle of the ladder, and they were jerking up and down, trying to get loose. Finally, half of the aileron was torn off the lower ship and I climbed up.

Other stungs that I used to do? I'd go out and hang by my toes through the cross wires. Then I had a trapeze that I attached to the landing gear with a mouthpiece on it, and I used to go down and hang from that by my teeth.

When I made a plane change, I used steel tubing rings and steel cables. One time we had thunderstorm activity, and flying around we accumulated a lot of static. So the minute I made contact, I got a terrific shock and almost let go. That was a very close shave.

What does the stick do? What's the rudder for? Such basic questions have produced surprising and controversial answers from America's foremost writer on aviation technique. What has made Langewiesche's writings endure for nearly fifty years, however, is neither his technical expertise nor his unconventional views, but his soul. He speaks to us in the fatherly manner of Alec Guinness explaining the Force in *Star Wars*. Through such books as *I'll Take the High Road* and *Stick and Rudder*, and his articles in *Air Facts* and *Flying*, he has been a guru to hundreds of thousands of pilots. The piece below is from *A Flyer's World*.

WOLFGANG LANGEWIESCHE

The Three Secrets of Human Flight

Sometimes I watch myself fly. For in the history of human flight it is not yet so very late; and a man may still wonder once in a while and ask: how is it that I, poor earth-habituated animal, can fly?

Any young boy can nowadays explain human flight—mechanistically: " ... and to climb, you shove the throttle all the way forward and pull back just a little on the stick. . . ." One might as well explain music by saying that the farther over to the right you hit the piano the higher it will sound. The makings of flight are not in the levers, wheels, and pedals but in the nervous system of the pilot: physical sensations, bits of textbook, deep-rooted instincts, burnt-child memories of trouble aloft, hangar talk. As I watch myself fly I think I can distinguish in these things some sort of order; I think they all add up to three basic mental adjustments that make the earth-habituated animal at home in the air: the three secrets, as it were, of the art of flying.

The first one is concerned with wings and how they fly: an understanding in your brain, and also a sort of wisdom in your hands until even your hands become as familiar with Bernoulli's theorem concerning the behavior of air as they are with Newton's law concerning the behavior of apples. Airport speech puts it that a fellow has got to know what an airplane will do: this seemingly fatuous statement has a lot of meaning; for airplanes, as any pilot will tell you, will do the damnedest things. When you throw an apple upward and let it fall into your hands again, the mechanics of the thing, though actually highly involved, seem "natural"; you understand them without even using your brain. But when you go riding around in the air on a wing, things happen which might make you think that the air is a crazy annex of our physical world in which all mechanical common sense is turned upside down.

The very beginning of aerodynamics — the science of lift — sounded a keynote. Magnus in 1852 investigated the behavior, then unexplained, of the spinning tennis ball which does not obey the laws of ballistics but flies along unexpected curves: in plain American speech, the screwball. The "Magnus effect" is the force exerted by the air on a ball that spins as it flies. It forces the ball off sidewise from its "proper" path. Well, in a very highbrow view, this effect is the same thing as the lift effect on a wing that holds an airplane up, when "properly," being heavy, it ought to fall down. The screwball flavor remains. Airplanes don't behave as they "should."

On wings it is safe to go fast, dangerous to go slow. It is safe to be high, dangerous to be low. If you want more lift, you pull your ship into a slightly more nose-high attitude; but pull it still a little higher, and your wings suddenly lose all their lift in a "stall" and let you down hard. When flyers want to say: "Take good care of yourself and don't fall down," they don't say as might be common sense, "Keep your nose up"; they say, "Keep your nose down." To make an airplane sit down on a runway and stay down, you pull the stick back as for an extreme upward zoom.

Take the airport story of how the great Eddie Stinson, out in Texas in 1916, discovered how to get out of a tailspin. It's probably not true, but it's well-invented. (Incidentally, the recovery from the spin was independently discovered by many different pilots.) The story is that Stinson had got into a spin. No man had ever recovered from a spin. You could pull back on the stick with all your might — nothing would make a ship's nose come up out of that twisting, uncontrollable corkscrew descent: a tailspin was death itself. Stinson, a tough man and one of the great pioneer flyers, thought he might as well get dying over with quickly. He pushed the stick forward for a steep dive, to get down faster and bash himself all the harder against the ground. The spinning stopped and the ship recovered. The suicidal reaction was the saving one!

This spin recovery is now a standard maneuver. Everybody learns it. And now that we understand it, it is not as fierce as it sounds. But still it goes contrary to ground-trained common sense. And it is still true in many flying situations that your ground-trained common sense doesn't work right. Your instinct of self-preservation would lead you into danger. The thing that seems right is wrong, the thing that is right seems unnatural. The prime example is the glide down to a landing. It's always assumed in flying that an engine can quit. Therefore, you must be able to reach a good field without help of the engine. Well, the field is pretty far away. You haven't much altitude left: can you reach it?

As the ground comes up relentlessly, the temptation is almost irresistible for a young pilot to pull back a little more on the stick, hold the nose a little higher, steer away from the ground, try to squeeze out a few more moments of flight, to "stretch the glide" another hundred yards or so that would get him across those trees into the cow pasture beyond. It is called the death glide. It doesn't work. The higher he holds his nose, the more the airplane sinks. The more the airplane sinks, the more he pulls back on the stick—and the result is a stall and nose dive into the ground. Around the old-time training fields they knew it well. They would sound the siren and jump on the crash truck before the student even realized he was in serious trouble.

The more logical your mind, the more exasperating is this contrariness of the rules of flying. An instructor friend of mine lost a customer over it, a woman lawyer who wanted to be taught to fly. Her trained sense of A and B was outraged by her first few hours of instruction. (At airports even men lawyers have a bad reputation as students: too logical). She posted on his hangar bulletin board a transcript of his patter in flight. It ran something like this:

"It's all very simple, see? Now you follow me through and take off. We want to go up, see? So first of all we get our tail up. We want to keep her straight, see? So we put some pressure on the right rudder. Now we gotta climb out over those trees, see? Well, don't hold your nose so high, you'll never make it that way.... Now cut your gun and do a landing.... Come on, pull back on that stick. Don't you see we are too high? Hold that nose a little higher. See, now we are coming down...." That's what she wrote, and then she left to get her recreation out of the logical and predictable flight of golf balls.

Eventually this whole screwball logic becomes clear to you. Just as there is sense in selling things on the exchange that you have not got, and buying things that you do not want, so there is sense in the apparent nonsense of aerodynamics. Just as the businessman's mind can translate inventory into cash, future earnings into present values, pesos into pounds sterling, and see the dollar sign in all of them, so does the pilot's mind translate speed into lift, lift into height, height back again into speed, and see them as different forms of the same thing. But it takes time for your brain to discover that. It takes more time for your body, your nerves, your instincts to get used to it.

After all, these things seem so contrary to common sense mostly because people have not worked much with the air before, except for a few centuries of inept fumbling with sails, kites, and windmills. Even the word airfoil is new. It means a body designed to grab useful forces out of flowing air; a sail or an airplane wing or a windmill blade is an airfoil. "Air flow" and "streamline," too, are new words and only half understood. Until quite recently, people would not believe that air offers a serious resistance to an automobile. People would not believe that the best shape to cut through air was blunt-nosed, with a long tapering afterbody; they wanted a sharp, pointed front such as would be right for a snowplow.

The screwball laws of the air are slow to become reasonable to our minds and even slower to become natural to our bodies, our nerves, our "feel." There is a nervous trouble among pilots called "aeroneurosis." It makes you irritable. It gives you stomach trouble. I think it is the fear of falling, firmly suppressed, popping up in disguise. I think it comes from the notion that you are really sitting on Nothing and that it ain't right. You cannot swim until you believe that the water will carry you. You cannot *really* fly until you believe that the air isn't a Nothing, but has substance and mass and the power to hold you up. My own last tenseness went only after I once made a delayed-opening parachute jump. I fell free for one thousand feet or so. The air, instead of being a void, turned out to be something on which you could lie as on a bed; something that you could feel yourself heavy against even as you fell.

Air and how it feels to a fast-moving body: that is perhaps the basic fact of human flight. If I had a boy who wanted to become a pilot, I would not prepare him by teaching him how to drive. Wheels and wings have nothing in common. Nor would I make him balance in high places; it's got nothing to do with flying. I would not teach him all about engines. I would certainly not teach him how to screw up his courage. I would give him a feel for the air by getting him to play with a piece of cardboard held outside an automobile window at fair speed. Even now I like to hold my hand out there and feel how the speed-stiffened air presses against it and pushes and pulls as I form it into different shapes and tilt it at different angles. The push and suction of fast-flowing air — that's what your wings feel. If you have felt it, you have already some understanding of wings.

Of all the amazing things about the airplane, the most amazing to the novice is that it has a will of its own. It *wants* to fly. Most of the time, the best thing you can do is to fold your arms, lean back, and let it fly. It may wallow in the gusts, but it will not capsize; it rights itself without help from your hand. Some air disturbance may throw it into a zoom or a shallow dive; then it does a couple of roller coasters, swinging back and forth between dives and climbs, but each one will be shallower than the last one, and eventually it will barge levelly ahead. It *wants* to fly. It hangs in the thin air as firmly and stably as a boat floats on the water.

The speech of American pilots expresses that. The airplane has had all sorts of names: "aerodrome" — this left the nature of its motion conveniently Greek and undefined; "car" — this irritated flyers because it suggested the wrong sort of motion; "machine" — but this does not reflect the almost-living behavior of the thing, its will to fly; "crate" — this became obsolete long ago, when steel-tube construction replaced wood, and airplanes no longer splintered apart in minor crack-ups. In the end, pilots have chosen the word "ship." This is strictly wrong English: a ship floats because it is light. In aeronautics, the word ship properly means a balloon-type aircraft. But they are right just the same. An airplane behaves under you like a ship, a broad-beamed, stable one. If you have ever held the tiller of a sailboat you already know the "control touch" of an airplane. A sailboat at times requires no steering at all but sails itself — because of the way things balance — the pressure of its sails, the weight of its keel, the shape of its underwater body. The airplane, too, flies itself, because of a similar balance — because of the place where its center of gravity is; because of the place where its wings are attached, and the exact angle at which they are set; because of its tail fins.

There is a newspaper phrase that pilot so-and-so "winged his way." It is misleading. It would make you think that a pilot keeps himself from falling down by some *knack*, by some continual fancy-business with his hands and feet. Actually a pilot no more "flies" his airplane than a sea captain "floats" his ship. The airplane flies itself. It happens almost every year, somewhere in the world, that an airplane, left unattended with the engine running, gets away pilotless. In most cases it rolls all over the airport, chasing its tail; sometimes it staggers aloft and crashes into a tree. But some have cleared all obstructions and flown successfully for hours, circling until their tanks were dry. When the engine stops, an airplane does not fall out of control; it simply noses down and glides. Some lucky ones came down on flat open fields and landed themselves undamaged. One runaway had sense enough to come home to its own airport for the landing! But it rolled into the airport fence.

But — this is my point — you have to believe that before it does you any good; believe it not only with your brain, but with your nerves as well. At first, it is sometimes harder to trust one's ship than to try to fly it. For the air is at first an odd place to be, one that *seems* to call for quick courageous action. I can see this idea at work when I take a newcomer up for a first try at the controls. I watch him in the rearview mirror; I feel his action by keeping my fingertips and toes on the dual control. There is the fellow who grits his teeth. He deals out quick thrusts in all directions as if he were fencing a duel against gravity. He has heard that a pilot needs "presence of mind." He is not going to let gravity get ahead of him if he can help it! There is the fellow who hardly dares to move the controls at all. He immediately counteracts any control impulse by its opposite, thus paralyzing himself; he keeps shoulders and elbows drawn in tight. He hardly even dares shift his glance away from the ship's center line. He has heard that a pilot needs a "keen sense of balance," and he behaves as if he were tight-rope-walking across a deep chasm.

Perhaps the greatest source of accidents in flying is excessive maneuvering; the itch for control; the craving to do something, born of the feeling that merely being up in the air calls for brave, energetic action. The art of flying lies often in *not* trying any tricks. A "foolproof" airplane, as pilots understand that term, is an airplane whose controls are restricted in motion. When you try to pull the stick too far back or kick a rudder pedal too far, you hit an iron stop. The most valuable item I find in my own bag of piloting tricks is a negative one. An old-timer taught it to me. I lift my hands and feet off the controls for a moment, especially in a difficult tight maneuver — say in a steeply curved landing approach. That keeps me from tensing up on the controls and squeezing my ship into a cramped condition of flight that might throw it out of control. One's trust in the airplane, one's trust in the air itself must be deep-seated and physical, not merely intellectual.

So, if you want to know what a pilot does as he "wings his way" across the country, imagine him doing nothing much; certainly nothing acrobatic or fast. He does much what a sea captain does on the high seas, who steers a course and keeps a lookout but otherwise simply bears the grandiose boredom of the sea.

The second of the three secrets of flight has to do with freedom of motion. The human organism must uproot itself, as it were, from the solid and stationary ground. The airplane is not merely fast; it is free to the sixth and last degree of freedom. That is, it can execute all the six conceivable kinds of motion all at once: speed forward, slip sidewise, climb or sink, roll over sidewise, pitch nose-up or nose-down, and yaw its nose to the right or left. But the pilot's own brain, his nerves, his body are at first still rooted to the ground.

Let some new, green flyer get into some sort of trouble up there alone, in the slippery, wallowy air, and the first thing his earth-habituated organism craves is something to grasp hold of, something solid. And he may use the control stick for a handhold! This is what is meant by "freezing to the stick." In the older days, when the air was even newer, and new flyers were even less nonchalant than now, it used to happen a lot, even during dual instruction. There were stories about how you had to have your little fire extinguisher handy — not to put out fires, but to knock out a student who froze on the controls.

Fast, free motion can have strange results. For a pilot, it must become reasonable and natural that heavy becomes light, and light heavy: in a steeply banked tight curve, your own bodily weight becomes 300, 500, even 1,000 pounds, as centrifugal force pulls you down into your seat. The blood inside you is pulled downward; your feet heat up. Your vision turns gray; your brain fuzzy. It is what one might call a power faint, an enforced bloodlessness of the brain. Your cheeks sag; your eyeballs flatten in their sockets; your muscular coordination, not attuned to such gravity, goes sour. You reach perhaps for the throttle, but your hand, which now weighs twenty pounds instead of five, arrives ten inches below the mark.

Curve your flight path some other way, and your weight becomes nothing at all. This is a good practical joke to play on a photographer. You lift him a few inches out of his seat when both his hands are full of camera. You keep him floating, weightless, dismayed, and pawing the air for a long moment. All it takes is a small gesture of the hand, a push on the stick. You drop the ship away from under him faster than he can fall, but not so fast that the cabin roof hits him on the head — though even that would be easy enough.

Curve your flight path still another way, and other startling things happen. The principle of acrobatic flying is that merely by curving your flight path you can command centrifugal forces that will paste you against the sky in any conceivable attitude whatsoever, at least for a moment. In the liberation of flight, up can become down. On top of a loop, for instance, it is reasonable and not supposed to be alarming to see the blue sky under you and the green earth hanging over you as a ceiling. Right and left can become top and bottom, as in a steep curve, when the ground is wheeling as a huge disk off your wing tip. In such a curve to change your heading from east through north to west means, for the flyer's physical sensations and as far as control action is concerned, not to turn, but to climb: you climb around in a horizontal loop.

That is why pilots find a loop, a spin, a slow roll once in a while "relaxing." It tears asunder the adhesions between the flyer's mind and the ground. It kills the idea—so strong in the beginner's mind and so easily re-formed even in the experienced pilot— that it is wrong to hold the ground in any attitude but level below him. Practical flying does not call for turning yourself upside down; but you do sometimes have to let the ground tilt up and wheel around in crazy arcs—and you have to feel at ease while it tilts and wheels. Every turn in an airplane, to name just one instance, involves a bank, much as a bicycle turns by banking. In a sharp, hurried turn you go way over. Such a turn may become necessary in an emergency close to the ground. And then the ground, vividly near, reasserts its hold on the pilot's mind, makes his control action timid and cramped, makes him reluctant to stick his wing tip down at the ground as steeply as required. "Flat," skidded turns lead to loss of control. They rank high among the causes of fatal accidents. "Ground-shy," says the airport postmortem.

Or consider how a pilot's mind functions while flying in a wind; how it must again tear itself free of the ground or else be tricked by optical illusion. What matters in flying, what keeps you up, is not motion pure and simple; it is motion *through the air:* the actual impact of your wings on particles of air. But—here is the pilot's problem— you cannot see the air; and therefore, you cannot see your motion through it. What you *can* see, what gives you a vivid sense of motion, is the ground. It slides along under you. But when you fly in a wind, the sliding of the ground is deceptive; it is then a compound motion, registering not only your progress *through* the air (which gives lift to your wings and is genuine flying motion), but also your drift *with* the air (which your wings cannot feel and which to the pilot's mind is therefore fake motion). Gauged by the ground, this fake motion adds itself to your every maneuver: the wind distorts your curves and makes you slide sideways from your straightaways. It slows you down when it is against you. It gives you fake speed when it is with you. Yet all the time you may be flying evenly and well-balanced through the air. In the air you may fly a continuous perfect circle, so perfect that you hit your own propeller wash every time: the ground registers a pattern that looks like a coil spring pulled out of shape.

In a fast wind with a slow airplane, these drift effects can disconcerting: I have actually moved backward over the ground while climbing a light airplane at 50 miles per hour into a strong wind. I have sometimes been swept sideways across the landscape nearly as fast as I was flying forward; "crabbing" is what pilots call that compound motion. It is much like what happens to a ferryboat crossing a swift river, except that in flying only the riverbanks are visible and the river not.

If you believe the ground, you get the urge to correct your maneuvers for looks; you want to make those skewed curves look safer, saner, not so drunken. Your ideas of how a vehicle should behave were formed in automobiles on concrete roads. You don't like that feeling of your being swept sidewise — and swept by a mere wind. But the pilot who gives in to terrestrial appearance spoils the aerial essence: the evenness of his motion through the air. You correct your curves for distortion, and find yourself skidding and slipping through the air; you try to stop the crabbing, and you find yourself crowding your ship with your controls "crossed," that is, mutually opposing each other. You believe the fake impression of speed as you fly downwind, and you come to in an ambulance.

If all this seems a bit abstruse and overburdened with logic, that is as it should be. Pilots themselves, who live in it, have much trouble grasping the more intricate phases of it and making them seem reasonable and natural. You ought to hear some of those hangar arguments! A fly is flying free inside a glass jar. Now, if the glass jar is moved about by hand what does the fly see, feel, and think? How would it maneuver? The theory is tricky and a matter of mathematics; but the practice is simple and a matter of nervous attitude, of abandon: that the flyer must let go of his nervous hold on the ground and abandon himself to the air; that he must think of himself, not as part of the landscape, but as part of the wind that blows across it.

The third of the secrets of human flight has to do with the senses. You develop a new eye, a new ear, a new touch; and that adds up into a sort of special air sense by which you know your way around.

When the terrestrial man first goes aloft he is deaf, blind, and numb. He has violent impressions; he sees faraway landscapes move in a strange manner, but his senses fail to deliver the information he needs — where he is, where he is going, and whether there is danger. As for his eye, it cannot see any of the essentials of flight: the air itself that bears him is invisible. The two big spiraling vortexes that his wings leave as their wake — invisible. The burbling and eddying of the air flow that occurs when he overpulls himself and stalls his wing: all invisible. "If we could only see the spray!" once mused the late Anthony Fokker wistfully.

Because the air looks like nothing, the flyer's eye has no perspective and no middle ground. Judgment of depth and and distance is labored and brain-interpreted, instead of vivid and immediate. A passenger doesn't get dizzy, although a stepladder would give him the jitters. A first-flighter sometimes thinks the airplane is standing still! And the fledgling pilot spends about half of his training time at one single problem: how to glide down for a landing and really hit the spot he is aiming at. His first attempts are like those of a drunken man trying to put a key into a keyhole. He overshoots or undershoots by half a mile.

As for the newcomer's ear, all it conveys to him is alarm. Ever since God equipped the lion with a roar, loud noise has been a frightening device of nature that bypasses all reason and attacks the soul direct.

Inside a cloud the pilot's natural sense equipment fails completely. With vision, the master sense, blotted out, all other senses are adrift and their indications meaningless: a loop, a curve, level flight, a spiraling dive — all feel much alike.

An old airport story tells of an army flyer going through high, thin, milky haze. He caught up with a slow civilian ship, also groping along there almost out of sight of ground. Proud of his speed, the lieutenant turned himself upside down and overtook the other in this position. So the civilian, distrustful of his vague impression that the earth was beneath him and trustful of the new evidence presented to him, promptly turned himself upside down too! This is not true, but it makes a point.

Flying "on instruments" — when you can't see anything — is a fierce and monkish art; a castigation of the flesh. You must cut out your imagination and not fly an airplane but regulate a half-dozen instruments — keep them in line, systematically slam, kick, or coax them to show what you want them to show, by using your rudder pedals and stick and throttle "on" them — coldly: forget about the airplane; never mind how it feels. Keep the instruments in line and the airplane will take care of itself. You must hold at bay all your natural senses, ignore your sensations of speed, of direction, of equilibrium. If you feel that you are hanging by your ear, but the instruments say you are level, you must not give in. At first, the conflicts between animal sense and engineering brain are irresistibly strong. You give in to the animal sense every time, and your instructor has to take over. Even after hours of instrument experience it comes back: sudden attacks of vertigo, cramps of the space-consciousness, powerful, like seizures. I have seen a student fly "on instruments" when he was leaning way over sideways in his seat. He was giving in, for himself, to vertigo. But for the airplane he was not giving in. He was flying at straight and level. I jostled him with my shoulder and told him to nod his head hard. That broke his spell. The penalty for not believing your instruments is a spiral dive at high speed.

Most flying is not done on instruments, but by the natural senses — the flying instinct, it used to be called. How does this air sense work? It is, of course, not a new sense at all, much less an instinct. It is merely ordinary sense perception doing new jobs; or perhaps ordinary sensory jobs, done by new and unexpected parts of the nervous system. "Now I get it!" cries the student while he is learning to fly a turn. "I get it by the seat of my pants!" And from then on he suddenly knows how to use his rudder pedals and his stick together to make a good turn: just use enough rudder to keep him sitting in his seat without sideways pull, as in an automobile on a banked curve. Though all the pilot's information is not gathered that way, most other ways of sensing are similarly unorthodox.

Of all possible maneuvers, take a gliding approach to a landing in a small field. Of all the pilot's sensing tasks during the glide, take only one: how he senses his speed. It is a vital problem. A couple of miles per hour too slow, and the wings stall; the ship drops out from under him, at an altitude too low to recover. A few miles per hour too fast, and the ship "floats" too long during the actual landing; it refuses to "sit down" and floats or rolls across the field into a fence. A delicate job of piloting. Vision is useless for it, because the wind would give you a fake impression of speed. The airspeed indicator is not sensitive enough. It is up to the pilot's senses.

In flying, speed can't be seen, but it can be heard. The flow of air past a ship's skin makes a hissing sound. When that sound is brisk, you don't worry. But when it peters out, you perk up: quick now, point your nose down a bit more, pick up more speed. Sailplanes, whose only sound it is, are flown largely by their hiss. In the old-style training ships — biplanes with lots of bracing wires — the glide was made mostly by the low humming of those wires. If you slowed such a ship up too much and pulled yourself closer and closer to a stall, the wires would hum down a descending melody: "Nearer, My God, to Thee." Or so 'twas said. When you get really acquainted with a ship, you often discover some warning sound that comes (not by design but by chance) just before the stall: a howl or a whistle or a rattle. In one ship I used to fly, the burbling of the airflow would set up a flutter in the fabric covering; it sounded as if someone were rapping sharply on the cabin roof: "Hey, you, that's about *It.*"

If such clues seem a bit flimsy, here is another: you can *feel* speed. Count Zeppelin, the airship master, once said that man would never fly safely on wings. Unlike the birds, man had no nerves in his wings. But the practiced pilot's sense of touch reaches all the way out to his wing tips. He feels the air flow out there much as one might hold out one's hand from a drifting canoe and feel the flow of the water. As he moves his stick, or wheel, and sends forces out to his wing tips, moving control surfaces there, the control surfaces send forces back to his hand. In brisk flight they send back a stiff, elastic resistance against his hand. In slow flight, the airflow slackens, and that resistance turns soft; the stick feels mushy. In some ships this control feel "goes out" completely when the wing begins to stall, and the stick moves loosely as if the wires had snapped.

Richard E. Byrd tells about that. On the way to the South Pole, his big trimotor, heavily loaded, was in trouble in the mountains; it would not climb any higher; when pointed up it threatened to stall. The pilot, Bernt Balchen, merely leaned back and showed Byrd and the others one thing: he could spin the big control wheel round with one finger. All in the ship understood what that means: quit trying to climb — or else the airplane would stall or spin. They jumped to the trapdoor and dumped some of their emergency supplies on the glacier below and eased their ship — and they made it. So there is another way to sense that all-important speed-through-the-air that you can't see: you *feel* for it, by small stirring motions of the controls, much as you might feel with your spoon for a lump of sugar in your coffee.

A pilot has a *feel* for speed in still another way. This one is the nearest thing to the mysterious Flying Instinct which used to be considered the thing that made a man a pilot. Flight surgeons call this kinesthesia; they speak of "joint and tendon sensing," "deep muscle sensing" — which can be loosely described as one's sensing of one's own weight, and the way you know what your own hands and feet are doing. Here is how it works: by the natural law of wings, speed and lift are two forms of the same thing. If you glide briskly and pull your nose up, you get a lift, and your ship comes up against you from underneath, like a horse taking off for a jump, and makes you feel heavy. If you glide slowly and try to pull some lift out of your ship, you get no response. And if your flying speed is almost gone and you try the same thing, your wings stall, the ship begins to settle away from under you, and you feel a softening and lightening of your seat. Take that process in extreme miniature: minute pressures of the hand on the controls, minute responses of the ship, minute variations of your sense of weight: in a "born" pilot they are blended into a continuous feel. "How do you know when you're about to stall?" I once asked Al Bennett, an instinctive flyer if there ever was one, and a man who taught me a lot. He said, "Why — it's all *through* me."

An intricate art? Or just a fancy way of driving a car? You can see how the popular ideas regarding human flight strike one who flies: They are almost all true, but seldom in the sense in which they are believed. Surely flying is as easy as driving a car — once you have understood how a wing flies. Surely it involves mechanics, but they are those of the airfoil, not of the garage. A housewife has as good a grounding in them as a tinkerer with machines: both start at scratch. Surely you need keen senses — but the sharpest vision won't help if you are fooled by wind drift. It is an intricate art but a teachable one. You need nerve, but only so long as you allow yourself to be puzzled by the air. Surely the air is full of mysteries and an eerie place to invade. But like the sea, it is friendly to those who are willing to be shaped by it.

This is a section of Leonardo's "Treatise on the Flight of Birds" *(Sul Volo degli Uccelli)*, translated by Edward McCurdy from the *Notebooks*. The powers of observation are awesome, the detail exquisite—but technologically, Leonardo's deductions were a dead end, both for himself and for those who followed slavishly in his imitation of nature; muscular power was beside the point. See the discussion by Lilienthal which follows and that on Leonardo's theories of flight by Parsons.

LEONARDO DA VINCI

The Flight of Birds

Write of swimming underwater and you will have the flight of the bird through the air. There is a suitable place there where the mills discharge into the Arno, by the falls of Ponte Rubaconte.

There are two different ways in which a bird can turn in any direction while continually beating its wings. The first of these is when at the same time it moves one wing more rapidly downwards than the other with an equal degree of force, the movement approximating toward the tail; the second is when in the same space of time the movement of one wing is longer than that of the other. Also in striking with the wings downwards slantwise, if they become bent or moved one lower down and the other farther back, the part which drives the wing lower down will be higher in the first case, and the opposite part of the wings will be higher in the first case, and the opposite part of the wings which has the longer movement backward will go farther forward through this first; consequently for this reason the movement of the bird will form a curve round that part of it which is highest.

These then are all the movements made by the bird without beating its wings, and they are each and all subject to a single rule, for all these movements rise upon the wind, for they expose themselves to it slantwise receiving it under their wings after the manner of a wedge.

A bird is an instrument working according to mathematical law, which instrument it is within the capacity of man to reproduce with all its movements, but not with a corresponding degree of strength, though it is deficient only in the power of maintaining equilibrium. We may therefore say that such an instrument constructed by man is lacking in nothing except the life of the bird, and this life must needs be supplied by that of man.

The life which resides in the bird's members will without doubt better conform to their needs than will that of man which is separated from them, and especially in the almost imperceptible movements which preserve equilibrium. But since we see that the bird is equipped for many obvious varieties of movements, we are able from this experience to declare that the most rudimentary of these movements will be capable of being comprehended by man's understanding; and that he will to a great extent be able to provide against the destruction of that instrument of which he has himself become the living principle and the propeller.

If a man have a tent made of linen of which the apertures have all been stopped up, and it be twelve braccia across and twelve in depth, he will be able to throw himself down from any great height without sustaining any injury.

You will perhaps say that the sinews and muscles of a bird are incomparably more powerful than those of man. . . .

But the reply to this is that such great strength gives it a reserve of power beyond what it ordinarily uses to support itself on its wings, since it is necessary for it whenever it may so desire to double or triple its rate of speed in order to escape from its pursuer or to follow its prey . . . and in addition to this to carry through the air in its talons a weight corresponding to its own weight. . . .

Man is also possessed of a greater amount of strength in his legs than is required by his weight. . . .

Observe how the beating of its wings against the air suffices to bear up the weight of the eagle in the highly rarefied air which borders on the fiery element! Observe also how the air moving over the sea, beaten back by the bellying sails, causes the heavily laden ship to glide onwards!

By adducing and expounding the reasons of these things you may be able to realize that man when he has great wings attached to him, by exerting his strength against the resistance of the air and conquering it, is enabled to subdue it and to raise himself upon it.

Otto Lilienthal's great advances in glider construction and techniques were famous not only in his native Germany; they fired the imagination of such men as Octave Chanute, Percy Pilcher, and the Wrights. The visionary passage below is from an essay in the Berlin journal *Prometheus*, published in 1895, a year before Lilienthal died from injuries sustained in a gliding accident. The translation is by Dorothy Waley Singer.

OTTO LILIENTHAL

The Problem of Flight

It has certainly not been made easy for human beings such as ourselves to traverse the realm of the air freely like a bird. But the longing to achieve this leaves us no repose; a single great bird circling above our head arouses in us the desire to soar through the firmament as he is doing.

The physical understanding of every ordinary person is sufficient for him to conjecture that only the correct key to this matter needs to be discovered in order that a completely new world of communication may be opened to us. How restfully and with what complete safety and amazingly simple means do we observe the bird gliding through the air. Can that not also be accomplished by man, with his intelligence and with the mechanical aids to power that have already enabled him to perform veritable miracles? And yet it is hard, extraordinarily hard, even to approach what Nature achieves so brilliantly. What immense efforts have been vainly exerted in the attempt to make the skill of the bird available to mankind. Science too has turned seriously to the question of flight. The phenomena of natural flight have been analyzed both anatomically and mechanically, optically by means of instantaneous photography and graphically by electrically recorded graphs. Now at last we have the bird in such a position that theoretically he cannot cheat us any more but in practice he still gives us the wrong result. As soon as we wish to apply our knowledge to actual flight, our clumsiness is deplorably obvious, and the swallows fly over our head laughing at us.

Perhaps there is no department of technology in which the pathway from theory to practice is so hard to find. . . .

Whether direct imitation of natural flight is one of many ways or is the only way to our goal is today still a matter of dispute. For example, to many technologists the flight of birds by means of wings appears too difficult to be practicable by machinery, and they would not dispense with the use of the propeller for air traction which has won such approval for work in water. But almost all of them agree that flight, if it is to be achieved, must be at high speed; and that brings us to a main difficulty in the discovery of [a method] for flight. . . .

Although the principle of the flight of birds has now been taken as a basis for most of these projects, in which forward-moving wing surfaces provide the lift, nevertheless the methods employed to imitate this natural flight mechanically are as numerous both as regards the production of the apparatus and the nature of the experiments as are the aeronautical engineers who are working on the problem, for each goes his own way in the matter. But all these individual ways lead as a rule to one and the same rock on which usually the idea, if not the ingenious flying machine itself, is shattered before it has had time to serve its purpose. Unfortunately progress is hardly ever made beyond the first attempt which usually ends either by failure to rise in the air or, if this is achieved, by inability to land with undamaged apparatus.

Everyone can easily picture to himself what it would mean, to fly through the air with the speed of an express train and then, to land safely without damaging the machine. But if such a feat be demanded from a large, heavy, and complicated machine, the prospect of a safe landing is all the less. It seems absolutely presumptuous to reckon on a happy ending to such a first attempt at flight, especially with complicated equipment.

If we could not convince ourselves every day of the ease and safety with which birds rush through the air and know how to move with the wind, we should positively despair of discovering the art of flight. But is there really a prospect that we can attain this skill? What then are actually the aims of the technology of flight? To what degree of perfection will it be possible to develop free human flight? Yes, *develop!* that is the right expression: *development* the right idea which we must take to heart in order to make progress in the technology of flight.

None can foresee today to what a degree of perfection man will succeed in developing free flight, because up to now far too little work has been done on this special development. If now and again any idea for a flying machine is carried out and it is shattered on the above mentioned rock, it will be of little significance for the development of dynamic flight. Moreover, work on flight is mostly pure theory, and that is of little help at the present stage of flight technology.

The theory of flight is today really no longer in too bad a state. Since we have elucidated the aerial resistance encountered by the bird's wing, and the economy of power effected by the curve of its profile, we can understand very well all the phenomena of natural flight. But what we must now develop is actual flight. We now have to eliminate purely practical difficulties, but these are greater than is [at first] apparent

Perfection cannot be achieved by violence. Just because the inventors of flying machines have usually demanded far too much at once from them, their positive success has been so small. To remain in the air without a balloon and to achieve free flight through the atmosphere are tasks which present such a new field of work that it will only be possible to become oriented in it gradually. He who discounts sound development through ever increasing experience of free, stable and safe movement through the air will get nowhere in this matter. . . .

After I had established that glider flights with quite simple apparatus can be carried out from elevated points with stability and safety over long distances in moderate winds, the next need was on the one hand to extend this gliding practice in ever stronger winds in order as far as possible to achieve the long flights that we so much admire in birds; and on the other hand to attempt to assist simple glider flight by dynamic means so as to lead gradually to continuous flight through stiller air. . . .

Also an apparatus for steering flight was soon attempted in practice. To produce the beating of the wings a machine driven by compressed CO_2 was used . . . For the rest, the handling of the apparatus is similar to that for simple glider flight, though the first timid attempts convinced me that if I had merely hurled myself straightway into the air with beating wings, the apparatus would probably not have landed without damage. New and unaccustomed phenomena are constantly appearing, and a single unlucky landing is enough to ruin the whole appliance. Here again it is a question of *not demanding too much at once*. I had therefore to resign myself at first to making only the usual glider flights with this larger and heavier apparatus which weighs 88¼ pounds, or twice as much as an ordinary glider, whereby I could practice certain and safe landing. Only now, when that stage has been successfully accomplished, can I cautiously attempt free flight with beating wings.

There may of course be other methods for the proper development of free human flight. But if so they will involve similar problems which must be dealt with in order to reach the solution of this difficult undertaking.

But whichever way you take, there is no hope of progress unless the experiments which are made permit instructive observation of a body moving in really free flight, for we are dealing here with quite new phenomena which we never encounter in other fields of technology. Stable, free flight in opposition to changes of wind, and safe landing from dynamic flight are factors for which at present very little practical experience is at our disposal, though they constitute the very essence of flight technology.

Nevertheless this circumstance only makes the solution of the problem of flight more difficult but by no means impossible. When it has been generally recognized in which direction flight research is needed, the forces now so scattered will be concentrated on the right objects for successful work on the steady development of free flight.

This is the concluding chapter of Anne Morrow Lindbergh's *North to the Orient* (1935), an account of a 1931 flight with her husband in which she served as radio operator and navigator. In a Lockheed Sirius, the Lindberghs flew from Maine across Canada to Alaska and Siberia, along the shores of the Arctic Sea, and thence to Tokyo and Nanking. An enormously popular book, *North to the Orient* sold 100,000 copies in its first six weeks and has remained in print almost constantly ever since. A contemporary reviewer credited its success to the author's "seeing eye and singing heart"; both are on display here.

ANNE MORROW LINDBERGH

The Glass-Bottomed Bucket

We were flying again, several years after our trip to the Orient. It was not a long flight nor an important one. It was not even particularly beautiful, just a casual trip from New York to Washington. We were not pressed for time; the weather was good; I had no radio to operate, no maps to look at. It was for me, simply flying, divorced from its usual accompanying responsibilities and associations. I could sit quite still and let the roar of the engine cover me like music. Throbbing with small monotone patterns, the vibration hummed in the soles of my feet, in the hollow of my back. It absorbed some restless side of me, and was satisfying as a hearth fire or rain on the roof. Contented, I could look down at that calm clear world below.

A new world, too, it was, for I had not flown in many months and the objects below me wore the freshly painted vividness of things seen for the first time. They passed, bright and irrelevant images, slowly under the still suspended wheel of our plane. (A wooded hill like moss, soft gray moss to crush in one's hand. The shadow of a single elm, flat on the ground, like a pressed fern. Pointed cedars and their shadows, two-pronged forks — for, in this world of flat surfaces, shadows are of equal importance with their objects. Pools in the fields as though the earth had just risen from the flood, shaking its shoulders. The sides of houses, hit by the morning sun, bright rectangles and squares, like the facets of cut stones.)

My eye, unaccustomed, temporarily, to such vast expanses to graze on, nibbled first here and then there at the scenes below, not finishing one patch in orderly fashion before starting on a new one. The images that attracted me were unrelated and scattered, not strung along one thread by a road, not cupped within the rim of a lake. (The pencil-marked shadows of telegraph poles. The neatly combed fields. Docks and piers and bridges, flat slabs laid on the edge of a mirror. Birds, particles of sand floating gently down the air. Cities, sudden flashes from an apartment window or a moving car — strange that the flash should reach such a distance, like a bright speck of glass in a road, sparkling far beyond its worth.)

There was no limit to what the eye could seize or what the mind hold — no limit, except that somewhat blurred but inescapable line of the horizon ahead. And even that line looked as though it might be limitless also. For if one swept the eye swiftly through the compass of the sky, one could see, or thought one saw, that slightly bowed look to the earth's surface. If I could turn quickly enough, I felt, I would catch sight of those flat fields and blue hills slipping down the side of the round world.

But still, I did not need to; there was too much to see as it was. A clear and perfect morning except for a slight ground haze, perceptible only in the distance, which, hanging over the earth, made a second horizon above the first one — as though the world were sunk under water, slowly emerging as the morning wore on.

Here below me I was not conscious of the morning haze. Through depths of clear transparent air, I looked down and saw those myriad bright shells on the floor of the sea. (Buoys newly painted and drying on a dock were scarlet lobster claws. Pierheads were pegs in the mud.) For the objects scattered below me bore no resemblance to those I had been living with. They bore no relation to life. Rootless and impermament, they seemed strewn there accidentally, washed up carelessly by some great tide of the sea; and left, limp, shining, detached, for me to pick up and arrange in what patterns I might choose.

They did, in fact, already form patterns (that strange blocklike pattern the rows of tenements made, doubled with their shadows; and the circular one of those black cars all centering to one point like an anthill), but they were new and different ones. They were patterns which seemed trivial and aimless from this great height, like the wavering, vinelike tracks crabs make in the sand. (How slowly those little cars crawl along the narrow ribbon paths!) And looking down on those little houses, those little paths, the narrow lines of black beetles, the anthill traffic of cars, one sat back and wondered, "Why? What do we do this for? Why isn't life simple and still and quiet? Was I really there yesterday? What was I doing?"

One could sit still and look at life from the air; that was it. And I was conscious again of the fundamental magic of flying, a miracle that has nothing to do with any of its practical purposes — purposes of speed, accessibility, and convenience — and will not change as they change. It is a magic that has more kinship with what one experiences standing in front of serene madonnas or listening to cool chorales, or even reading one of those clear passages in a book — so clear and so illuminating that one feels the writer has given the reader a glass-bottomed bucket with which to look through the ruffled surface of life far down to that still permanent world below.

For not only is life put in new patterns from the air, but it is somehow arrested, frozen into form. (The leaping hare is caught in a marble panel.) A glaze is put over life. There is no flaw, no crack in the surface; a still reservoir, no ripple on its face. Looking down from the air that morning, I felt that stillness rested like a light over the earth. The waterfalls seemed frozen solid; the tops of the trees were still; the river hardly stirred, a serpent gently moving under its shimmering skin. Everything was quiet: fields and trees and houses. What motion there was, took on a slow grace: the crawling cars, the rippling skin of the river, and birds drifting like petals down the air; like slow-motion pictures which catch the moment of outstretched beauty — a horse at the top of a jump — that one cannot see in life itself, so swiftly does it move.

And if flying, like a glass-bottomed bucket, can give you that vision, that seeing eye, which peers down to the still world below the choppy waves — it will always remain magic.

This piece is taken from *The Spirit of St. Louis*, Charles Lindbergh's Pulitzer Prize–winning book of 1953. It was his second book on the great transatlantic solo flight, the first one being *We* (the duo signified the man and the plane, a milestone metaphor of humanity's entrance into the machine age). That quite poor book was hurried out in 1927 to capitalize on the unprecedented fame which was instantly Lindy's upon landing at Le Bourget on May 21 of that year; this quite splendid book, the product of twenty-six years' rumination and reflection, does justice to the man and his feat.

CHARLES LINDBERGH

Flying Blind Over the Atlantic

THE NINETEENTH HOUR
Over the Atlantic
Time — 1:52 A.M.

WIND VELOCITY	*Unknown*
WIND DIRECTION	*Unknown*
TRUE COURSE	————
VARIATION	————
MAGNETIC COURSE	————
DEVIATION	————
COMPASS COURSE	————
DRIFT ANGLE	————
COMPASS HEADING	96°
CEILING	*Unlimited above clouds*
VISIBILITY	*Unlimited outside of clouds*
ALTITUDE	9,000 feet
AIR SPEED	87 *mph*
TACHOMETER	1,625 *rpm*
OIL TEMPERATURE	35° *C.*

OIL PRESSURE	59 *lbs.*
FUEL PRESSURE	3 *lbs.*
MIXTURE	4
FUEL TANK	*Nose*

Eighteen hundred miles behind. Eighteen hundred miles to go. Halfway to Paris. This is a point I planned on celebrating out here over the ocean as one might celebrate a birthday anniversary as a child. I've been looking forward to it for hours. It would be a time to eat a sandwich and take an extra swallow of water from the canteen. But now all this seems unimportant. Food, I definitely don't want. And water—I'm no longer thirsty; why trouble to take another drink? I have as far to go as I've come. I must fly for eighteen endless hours more, and still hold a reserve for weather. Time enough for food and water after the sun rises and I wake; time enough after the torture of dawn is past.

Shall I shift fuel tanks again? I've been running a long time on the fuselage tank. I put another pencil mark on the instrument board to register the eighteenth hour of fuel consumed. That wasn't so difficult; it didn't require any thought—just a straight line, a quarter inch long, one more in those groups of fives. But shall I shift tanks? Let's see; how did I plan to keep the balance? Oh, yes; it's best not to let the center of gravity move too far forward, so the plane won't dive under the surface in case of a forced landing. I turn on the nose tank, and shut off the flow from the fuselage tank, instinctively.

There's one more thing—the change of course—each hour it has to be done. But what difference do two or three degrees make when I'm letting the nose swing several times that much to one side or the other of my heading? And there are all the unknown errors of the night. Sometime I'll have to figure them out—make an estimate of my position. I should have done it before; I should do it now; but it's beyond my ability and resolution. Let the compass heading go for another hour. I can work it all out then. Let the sunrise come first; with it, new life will spring. My greatest goal now is to stay alive and pointed eastward until I reach the sunrise.

During the growth of morning twilight, I lose the sense of time. There are periods when it seems I'm flying through all space, through all eternity. Then the world, the plane, my whereabouts, assume unearthly values; life, consciousness, and thought are different things. Sometimes the hands of the clock stand still. Sometimes they leap ahead a quarter-hour at a glance. The clouds turn from green to gray, and from gray to red and gold. Then, on the thousandth or two thousandth time I'm leveling out my wings and bringing the nose back onto course, I realize that it's day. The last shade of night has left the sky. Clouds are dazzling in their whiteness, covering all of the ocean below, piled up in mountains at my side, and—that's why I've waked from my dazed complacency—towering, a sheer white wall ahead!

I have only time to pull myself together, concentrate on the instruments, and I'm in it—engulfed by the thick mist, covered with the diffused, uniform light which carries no direction and indicates no source. Mechanically, I hold my hand out into the slip-stream. The temperature of air is well above freezing—no danger from ice.

Flying blind requires more alertness. And since alertness is imperative, I find it possible to attain. I'm able to accomplish that for which there's no alternative, but nothing more. I can carry on the essentials of flight and life, but there's no excess for perfection. I fly with instinct, not with skill.

The turn-indicator must be kept in center. That's the most important thing. Then the airspeed needle must not be allowed to drop or climb too far. The ball in the bank-indicator can wait until last; it doesn't matter if one wing's a little low, as long as everything else is in position. And at the same time, I have to keep the earth-inductor needle somewhere near its lubber line. Thank God it's working again. Altitude isn't so important. A few hundred feet up or down makes little difference now.

The knowledge of what would happen if I let those needles get out of control does for me what no amount of resolution can. That knowledge has more effect on my mind and muscles than any quantity of exercise or determination. It compresses the three elements of existence together into a single human being.

Danger, when it's imminent and real, cuts like a rapier through the draperies of sleep. The compass may creep off ten degrees without drawing my attention; but let the turn-indicator move an eighth of an inch or the air speed change five miles an hour, and I react in an instant.

It's not a large cloud. Within fifteen minutes the mist ahead brightens and the *Spirit of St. Louis* bursts out into a great, blue-vaulted pocket of air. But there are clouds all around—stratus layers, one above another, merging here, separating there, with huge cumulus masses piercing through and rising far above. Sometimes I see down for thousands of feet through a gray-walled chasm. Sometimes I fly in a thin layer of clear air sandwiched between layers of cloud. Sometimes I cut across a sky valley surrounded by towering peaks of white. The ridges in front of me turn into blinding flame, as though the sun had sent its fiery gases earthward to burn away the night.

Another wall ahead. More blind flying. Out in the open again. But only for minutes. The clouds are thickening. I'm down to nine thousand feet. Should I climb back up where valleys are wider? No, I've got to get under these clouds where I can see waves and windstreaks. I *must* find out how much the wind has changed. I *must* take hold—begin to grapple with problems of navigation. The rising sun will bring strength—it *must!* Half the time, now, I'm flying blind.

When I leave a cloud, drowsiness advances; when I enter the next, it recedes. If I could sleep and wake refreshed, how extraordinary this world of mist would be. But now I only dimly appreciate, only partially realize. The love of flying, the beauty of sunrise, the solitude of the mid-Atlantic sky, are screened from my senses by opaque veils of sleep. All my remaining energy, all the attention I can bring to bear, must be concentrated on the task of simply passing through.

THE TWENTIETH HOUR
Over the Atlantic
TIME—2:52 A.M.

WIND VELOCITY	*Unknown*
WIND DIRECTION	*Unknown*
TRUE COURSE	——
VARIATION	——
MAGNETIC COURSE	——
DEVIATION	——
COMPASS COURSE	——
DRIFT ANGLE	——
COMPASS HEADING	96°
CEILING	*Flying between cloud layers*
VISIBILITY	*Variable*
ALTITUDE	8,800 *feet*
AIR SPEED	89 *mph*
TACHOMETER	1,625 *rpm*
OIL TEMPERATURE	35° C.
OIL PRESSURE	59 *lbs.*
FUEL PRESSURE	3 *lbs.*
MIXTURE	4
FUEL TANK	*Nose*

I change neither fuel tanks nor course.

This is morning—the time to descend and make contact with the ocean. I look down into the pit I'm crossing, to its misty gray bottom thousands of feet below. The bottom of that funnel can't be far above the waves. Then is the ocean covered with fog? Suppose I start down through these clouds, blind, where should I stop—at 2,000, at 1,500, at 1,000 feet? I reset my altimeter when I was flying close to the water, east of the New-foundland coast. But that's almost eight hours back, now. Since then I've crossed an area of major storm. The barometric pressure has surely changed during the night. How much, there's no way of telling. I think of the Canadian pilot, caught in fog, who flew his seaplane into the water without ever seeing it. I'd be taking a chance to descend below a thousand feet on my altimeter dial. It would be cutting the margin close to fly blind even at the indicated elevation. No, I'll hold my altitude a little longer. The climbing sun may burn a hole through the clouds.

As sky draws attention from the earth at night, earth regains it with the day. Sometimes unperceived, during this hour of morning twilight, I took back the earth and relinquished the sky. I no longer watch anxiously for stars in the heavens, but for waves on the sea. The height of cloud above is now less important than the depth of cloud below.

I've been tunneling by instruments through a tremendous cumulus mass. As I break out, a glaring valley lies across my path, miles in width, extending north and south as far as I can see. The sky is blue-white above, and the blinding fire of the sun itself has burst over the ridge ahead. I nose the *Spirit of St. Louis* down, losing altitude slowly, 200 feet or so a minute. At 8,000 feet, I level out, plumbing with my eyes the depth of each chasm I pass over. In the bottom of one of them, I see it, like a rare stone perceived among countless pebbles at your feet — a darker, deeper shade, a different texture — the ocean! Its surface is splotched with white and covered with ripples. Ripples from 8,000 feet! That means a heavy sea.

It's one of those moments when all the senses rise together, and realization snaps so acute and clear that seconds impress themselves with the strength of years on memory. It forms a picture with colors that will hold and lines that will stay sharp throughout the rest of life — the broad, sun-dazzled valley in the sky; the funnel's billowing walls; and deep down below, the hard, blue-gray scales of the ocean.

I nose down steeply, resetting my stabilizer as pressure on the stick increases. Controls tighten . . . ribs press against fabric on the wings . . . the airspeed needle rises . . . 110 . . . 120 . . . 140 miles an hour. I close the mixture control and pull the throttle back still farther, letting the engine turn just fast enough to keep it warm and clear. Air crowding around the cowlings screams strangely in my deafened ears, the first different sound I've heard since takeoff, yesterday.

A layer of cloud edges over the ocean. I turn sharply back to spiral through the open funnel. I forget about my plan to turn the altitude of night into distance during day. Those thousands of feet I've hoarded, I'll squander on the luxury of coming down with sight. Suppose I lose ten or fifteen miles in range. It's worth that to get down in safety to the wind-swept sea.

I bank again as another wall of cloud approaches. The shadow of my plane centers in a rainbow's circle, jumps from billow to billow as I spiral. . . . The sun's rays flood through the fuselage window, cut across my cockpit, touch first this instrument, then that. Whitecaps sparkle on distant water. . . . I'm banked steeply. . . . I'm descending fast. . . . My ears clear, and stop, and clear again from change in pressure. . . . My air cushion wilts until I feel the hard wicker weaving of the seat. . . . Wings flex in turbulence. . . . Layer after layer of thin gray clouds slip by, merging here, broken there; mountains, caverns, canyons in the air.

Two thousand feet now — under the lowest layer of clouds. The sea is fairly writhing beneath its skin — great waves — breakers — streaks of foam — a gale wind. From the

Charles Lindbergh

LUCKY **L**INDY EMERGED FROM HIS AIRPLANE BEFORE A CROWD OF 100,000 CHEERING FRENCHMEN AT LE BOURGET FIELD NEAR PARIS ON THE EVENING OF **MAY 21, 1927**. **H**E'D FLOWN 3,600 MILES NONSTOP FROM NEW YORK IN 33½ HOURS — **THE FIRST SOLO FLIGHT ACROSS THE ATLANTIC.**

THE WORLD SALUTED AMERICA'S HEROIC "LONE EAGLE." **H**E WAS AWARDED NUMEROUS EUROPEAN MEDALS, AND, UPON HIS RETURN, HE WAS OFFICIALLY WELCOMED BY PRESIDENT CALVIN COOLIDGE AND GIVEN THE DISTINGUISHED FLYING CROSS.

CITIZENS OF ST. LOUIS CONTRIBUTED FUNDS FOR LINDBERGH'S AIRPLANE, "**THE SPIRIT OF ST. LOUIS.**"

IT IS NOW ON DISPLAY IN THE NATIONAL AIR MUSEUM OF THE SMITHSONIAN.

northwest? I've been spiraling so long that I'm not certain of direction. I straighten out and take up compass course. Now I'll have the answer I wanted so badly through the night. . . . .I'm pointed obliquely with the waves. . . . Yes, the wind's northwest. . . . It's striking the *Spirit of St. Louis* at almost the same angle it blew off the coast of Newfoundland at dusk—but it's much stronger. A quartering tail wind! It's probably been blowing that way all night, pushing me along on my route, drifting me southward at the same time.

A tail wind! A tail wind across the ocean. That's what I've always wished for. How strong is it? I can judge better close to the surface. I ease the stick forward and begin a slow descent, translating my remaining altitude into extra miles toward Europe. The air's warmer and more humid—a different atmosphere than that above the clouds. It's like stepping through the door of a greenhouse full of plants.

I'm under a dark stratus layer of cloud. Only a spot of sun-brightened water behind marks the bottom of the funnel I spiraled through, as though the beam of a great searchlight had been thrown down from the heavens to guide me to the ocean's surface.

Curtains of fog hang down ahead and on each side, darkening the air and sea, shutting off the horizon. I nose down to 1,000 feet . . . to 500 . . . to 50 feet above huge and breaking waves. The wind's probably blowing 50 or 60 miles an hour. It would have to blow with great force to build up a sea like that—to scrape whitecaps off and carry the spray ahead like rain over the surface. The whole ocean is white, and covered with ragged stripes of foam.

It's a fierce, unfriendly sea—a sea that would batter the largest ocean liner. I feel naked above it, as though stripped of all protection, conscious of the terrific strength of the waves, of the thinness of cloth on my wings, of the dark turbulence of the storm clouds.

This would be a hellish place to land if the engine failed. Still, it wouldn't be as bad as a forced landing during the night—gliding blindly down through freezing mist and onto an ice-filled ocean. Nothing that could happen now would be as bad as that. Now, at least, I know which way the wind's blowing; I could head into it and stall onto the water with almost no forward speed at all. I could *see* what I was doing.

It would be awfully difficult to work down there with waves breaking over the fuselage and whipped by a gale of wind. The cockpit would probably fill up with water a few seconds after I landed. I might have to hold my breath until I could crawl out through the door and up on top of the plane. Then, I'd have to cut through the roof of the fuselage to get at the raft, and hang on to something while I pumped it up. After that, there'd be the problem of getting my equipment unlashed from the steel tubes under water, and getting it lashed again inside the raft.

Suppose I could get the raft pumped up and loaded. What then? While I was in San Diego planning the flight, I considered using the plane as a sea anchor and signal of distress if I were forced down in the Atlantic. The silver wings would be more likely than my small, black-rubber raft to attract the attention of any ship that passed.

But there's not likely to be a passing ship at this latitude. Looking down on the wilderness of broiling water, I realize that mooring my raft to the plane would cause the waves to break over it—if the cord held in such a sea. It would be better to cut loose and drift with the wind, southeast toward the ship lanes. At least then I'd be going somewhere. That would be preferable to waiting in one spot, watching an empty horizon, anchored to a sinking plane up in this northern ocean.

But there are other things more important than imagining forced landings, and now that I possess my senses, I must keep them disciplined. It's essential to take stock of my position, to lay out a definite plan of navigation for the day. The wind aloft is probably stronger than it is down here. If it also is from the northwest, and if it didn't shift during the night, I must be well ahead of schedule and south of my course. In that case, I should be about over the middle of the ocean.

The middle of the ocean! I glance down at the chart—somewhere in that empty space between the continents, somewhere among those small black numerals which represent mountains and valleys under sea—1,600 fathoms—2,070 fathoms—1,550 fathoms, the figures read. Yes, there's ground down below, just as contoured and distinctive as the ground of which continents are made. Water is like fog, hiding the earth from human eyes. If I could see through it, I might locate my exact position from some submarine mountain range.

What wouldn't I give for a high cloud's shadow on the surface to tell me the wind drift aloft! Now, whatever estimate I make is just a guess, a probability on which I can base—only hope. But right or wrong, I've got to make some estimate. What *should* I allow for the wind, for the swinging of the compasses, for those detours around thunderheads? And what bothers me still more, how shall I allow for the inaccuracy of my navigation—for those swerves to right and left of course during innumerable minutes of unawareness? When I left Newfoundland, I set my heading 10 degrees northward to compensate for drift. Should I now allow 5 degrees more? I look at the waves again. The wind streaks are really more tail than side—my route curved southward during the night. Fifteen degrees might be too much. Then for over an hour I haven't reset the compass at all. That leaves me headed an extra 2 degrees toward the north.

I have a strong feeling that I'm too far south to strike Ireland unless I change my heading. But there aren't enough facts to back it up. I must consider only the known elements in navigation. If I give way to feeling, that will remove all certainty from flight. Suppose I crank in 5 degrees to the earth-inductor compass, then if I don't make a landfall by. . . .

The waves ahead disappear. Fog covers the sea. I have only time to reset the altimeter and start climbing. A hundred feet above water, in rough air, is no place for blind flying. Turbulence is severe. The safety belt jerks against me. Needles jump back and forth over dials until I can follow only their average indication with controls. I push the throttle forward, and hold 95 miles an hour until the altimeter shows 1,000 feet. I watch the tachometer needle. It's steadier, and tells the position of my nose on the horizon more accurately than the inclinometer and the airspeed indicator put together. Problems of navigation fade into the immediate need of holding the *Spirit of St. Louis* level and on course. Adding and subtracting degrees, and keeping one result in my head while I consider some related factor, is too much. And to work with pencil and paper at the same time I watch those needles is out of the question. I'll figure it all out accurately after the fog has passed.

The fog doesn't pass. I go on and on through its white blankness. I'm growing accustomed to blind flying. I've done almost as much on this single trip as on all my flights before put together. Survival no longer requires such alertness. Minutes mass into a quarter-hour. A quarter becomes a half, then three-quarters. Still the waves don't appear. I'm flying automatically again through eyes which register but do not see. . . .

"The secret of flying," says the narrator, "is learning to minimize the risks" — and remembering always that man is an intruder in the air. Wise words from Gavin Lyall's 1966 suspense novel, *Shooting Script.*

GAVIN LYALL

Flying in the Face of Nature

I ate a couple of hot dogs on the gallery and was back at the Dove by two. Now the day was really coming to the boil. On the airport it was just plain hot, but across the bay in Kingston, sitting in its bowl between the hills, the air in the narrow streets would be like breathing under a sweaty electric blanket. Out on the box-top shacks of the nameless town built on the city dump up by the oil refinery, there would be sudden, vicious fights over a choice piece of rubbish. And up in the rich suburbs beyond Half Way Tree, dignified elderly gardeners would move listlessly among the mangoes, feeling the long sharp edges of their machetes and thinking of the owner's wife asleep in the air-conditioned room upstairs.

Kingston, the perfect natural harbor — except that it's on the south coast and the cool summer breeze comes from the north. So it only blows — no, it doesn't blow, just breathes politely — on the private beaches and modern houses and big hotels of the north coast. So move up there, man; nobody's stopping you. All you need is the money. And you think you're going to get rich in Kingston? Haul your rice, man; emigrate. Maybe London isn't paved with gold, but it's cool, man. Yes, you'll find out how cool.

Perhaps the heat was getting at me, too. The Dove was too hot to touch without gloves and there was a faint haze over the fuel-tank vents, so I was wasting petrol by evaporation. Well, I hadn't paid for it, anyway.

Customer number 3 was late. He always was, but he was still the only regular income I had. A young Venezuelan businessman named Diego Ingles who'd got the idea that his company would buy him a brand-new twin-engined airplane the moment he'd qualified to fly it. He hired me twice a week for twin-engine instruction.

Personally I had my doubts that his company could go crazy enough to hand him a 35,000-pound plane, but perhaps it could happen. He obviously came of a genuine aged-in-the-money family back in Caracas, and that counts for a lot in Venezuelan business circles. Anyway, it was his money and my pocket.

All that apart, he was a nice young lad: in his early twenties, shortish and slightly tubby, with a flat cheerful face, a bush of dark hair and the politely rakish manner of an old Spanish family upbringing.

He finally appeared at twenty past, with a long graceful apology which boiled down to the fact that he'd only just got out of bed, and not even his own.

With the heat and the tall thunderclouds building up on the Blue Mountains I wanted to get away from the airport, so we skipped the circuits-and-landings and I gave him a dead-reckoning navigation exercise out to the Pedro Cays, about eighty miles to the southwest. No radio to be used: he had to do it on maps and weather reports alone.

It wasn't his favorite type of flying: quiet, steady, accurate. Like most of Latin America he believed that aviation was a branch of sports-car racing. I'd had to keep taunting him about becoming a "fair-weather pilot" to keep him looking at—and believing—his instruments, maps and forecasts. This time I was feeling irritable enough to taunt him into making a near-perfect landfall over Northeast Cay within a minute of his ETA.

I remembered to tell him I'd noticed.

He smiled very charmingly and asked: "So perhaps you think I am good enough?"

I looked at him. "For what? You could probably get your license up-rated to a 'B,' so you could fly this size of thing privately. That's if you've been doing any book-work. D'you want me to arrange a test?"

"Not quite that, Señor. I mean—do you believe in me? Can I use an airplane like this—anywhere at any time?"

"Nobody can. You still think an airplane's a miracle with a starter button. Some weather, even the birds are walking."

"I understand there may be risks, Señor, but. . . ." He took his right hand off the wheel and fluttered it delicately.

"Just try and remember," I said slowly, "that if God had intended men to fly He'd have given us wings. So all flying is flying in the face of nature. It's unnatural, wicked and stuffed with risks all the time. The secret of flying is learning to minimize the risks."

"Or perhaps—the secret of life is to choose your risks?" He smiled disarmingly. "But I think you were a fighter pilot, and yet you talk of minimizing risks?"

"That's where I learned it. Don't fall for the King-Arthur-of-the-air stuff about fighter pilots. Clean knightly combats and all that. It's the one trade where the whole point is to catch a man by surprise and shoot him in the back. That's how the top men made their scores. And if they *couldn't* catch a man like that, they didn't tangle with him."

"Ah, now I know, Señor." He ducked his head gracefully. "So I should catch the weather by surprise and shoot it in the back. And also I see why the unromantic English make such good pilots."

He was laughing at me, but as long as he remembered. . . . I said: "So what about a license test?"

"I will talk to Caracas about it. But the important thing is that *you* believe I am good enough. That is my true examination." It takes generations of high Spanish blood to laugh at a man so courteously.

I put my pipe in my mouth and said: "You may still find a license is useful. In fact, licenses are useful in all walks of life."

"Señor?"

"Just an old saying the unromantic RAF had: only birds can fornicate *and* fly. And birds don't booze."

The thunderclouds had finally, mercifully, split and drenched Kingston for half an hour by the time we got back over Palisadoes. Steam was rising gently off the suburbs behind the town: even the air in the cockpit felt fresh and green.

We tried three landings, including one with an engine stopped and abandoning the approach at 200 feet and going round again. He handled it pretty well; he wasn't going to get killed in an emergency—that was his best time. It would be the small things that killed him: a bit of fluff in a carburetor and a slipped connection in a radio and a twenty-degree shift in wind—incredible coincidences of bad luck like that, in a time when the millions of flights every year make a million-to-one chance a tenfold statistical certainty.

He would die in a clear, still blue sky—because he still believed he had a *right* to be there. Because he wouldn't believe he was a trespasser who had to keep awake and alert every single damn minute.

On December 11, 1940, at the age of nineteen, John Gillespie Magee, Jr., was shot down over England. Born in Shanghai, the son of an American missionary, he was educated in England and in 1940 joined the Royal Canadian Air Force. This sonnet inspired Allied airmen throughout World War II and is Magee's legacy to aviators everywhere.

JOHN GILLESPIE MAGEE, JR.

High Flight

Oh, I have slipped the surly bonds of Earth
And danced the skies on laughter-silvered wings;
Sunward I've climbed and joined the tumbling mirth
Of sun-split clouds — and done a hundred things
You have not dreamed of — wheeled and soared and swung
High in the sunlit silence. Hov'ring there,
I've chased the shouting wind along and flung
My eager craft through footless halls of air.

Up, up the long, delirious burning blue
I've topped the wind-swept heights with easy grace,
Where never lark, or even eagle, flew;
And, while with silent, lifting mind I've trod
The high untrespassed sanctity of space,
Put out my hand, and touched the face of God.

Why do they become astronauts? Why climb Everest? The central idea of America may be that of the frontier, a land frontier which Theodore Roosevelt proclaimed closed just as the Wrights opened another. Once the atmosphere began closing in, outer space beckoned. No, they can't all live in the White House, and they can't all become household gods courtesy of television. Mailer suggests that the drive to become an astronaut may be not temporal but spiritual.

NORMAN MAILER

Why Do They Become Astronauts?

A man who becomes a good fighter pilot does not think anything can ground him. Fighters are like downhill racers — their sanity is that they do not look to pick up a sane option. Their balance is at the edge of balance.

Fighter pilots growing older sometimes become test pilots. The chances they used to take casually now collect around the act of flying the most dangerous aircraft to be found. But they have developed into engineers.

Once they become astronauts, however, they hardly test anything. Where once they might have been testing planes every two or three days, now they do not have more than one rocket flight every two or three years. They still fly, they fly T-38 jet trainers for hundreds of hours a year, but they are not testing the T-38, they are merely flying in order not to lose their reflexes. And astronaut training is in simulators or classrooms. They are on call for public appearances. As test pilots they flew rocket planes at four thousand miles an hour and lived as they pleased, free to carry a drinking party to the dawn or to search for solitude; now that they were astronauts they were obliged to live in homes of a certain price in suburbs of an impeccable predictability in a world of public relations where they were rendered subservient to propriety by a force of mysterious propriety within NASA itself, a force which might just as well have seeped from every door of every office until anything spontaneous in a man was stuffed into the cellar of his brain. They worked long hours, perhaps an average of fifty hours a week, and hardly knew when they would have their first space flight, or once up, whether they would ever have a space flight again. They were the best pilots in their profession, but now they flew only for practice, and they could not know for certain whether they would ever be able to practice their new profession. (It is as if Truman Capote gave up literature because he wished to write opera and suddenly could not find out whether any of his music would ever be sung.) All the while the astronauts were obliged to live in intense competition with one another, yet had to exhibit every face of good spirit and teamwork to the world. Stories were common at the Manned Spacecraft Center of astronauts who had shared the same flight yet hardly spoke to each other in the months before, so intense was their mutual dislike. Still they kept their animosity private for fear they could lose their seat on the flight. Then all of them had had to swallow the wrath they might have felt for the contractors connected to the fire which killed Grissom, Chaffee, and White. Say it worse. Eight of them altogether had already been killed. Besides the fire, four had gone in fatal crashes of the T-38, and one in an auto accident. (Astronauts were quietly famous for driving their cars a foot apart at a hundred miles an hour). One could say that the demand for order, hard work, and propriety in such competitive near-violent men had produced their deaths, as if the very tension of their existence as in a game of musical chairs had pushed the escape from death of one man over into a higher potentiality for accident of another. Being an astronaut was perhaps the most honored profession in the nation, but of a total of sixty-six astronauts accepted since the program began, eight were dead and eight had resigned. Since seventeen of the original sixty-six were scientist-astronauts, and only three of them had separated from NASA, it meant that thirteen of the forty-nine flying astronauts were no longer present. The resignation and mortality rates are not so close in other honored professions. Of course, most of the astronauts work for only thirteen thousand dollars a year in base pay. Not much for an honored profession. There are, of course, increments and insurance policies and collective benefits from the *Life* magazine contract, but few earn more than twenty thousand dollars a year.

So we are obliged to consider why a man would divorce himself from his true talent — which is to test a new jet or rocket plane — and live instead in propriety, order, competition, and tension for twenty thousand dollars a year, knowing he could make three to five times that much in private life, and not be afraid to utter a resounding opinion, get drunk in public, yell at his children before strangers, or be paralyzed by scandal or divorce. Can it be that any man who takes up such a life for thirteen thousand dollars base pay a year is either running for President, patriotic to the point of mania, or off on a mission whose root is the field of the magnet in the iron of the stars?

Some may have been running for President. John Glenn had been campaigning for senator before his accident in the tub, Borman was now close to Nixon, Schirra was a television commentator (a holding position) and Collins was yet to enter the State Department. And there were bound to be others. If an astronaut had political ambitions he did not necessarily announce them.

Then there were men for whom a celebrity as astronaut was preferable to the professional anonymity of the test pilot. And some were patriots. There is no need to diminish the power of this motive. Once, in a meeting of astronauts, NASA executives and scientist-astronauts, the NASA administrator, then James Webb, had told them there would be a hiatus in the Space Program during the early 1970s due to budget-cutting. The scientist-astronauts were gloomy. Last to arrive in the program, unscheduled for flights, they saw a delay of a decade or more before they could even go up. Scientific examination of the moon and space by experts such as themselves would be again and again delayed. One of the scientist-astronauts said, "Mr. Webb, this hiatus you've been referring to — how would you say that the scientific community — "

"To *hell* with the scientific community," Frank Borman cut in. The astronauts laughed. The attitude was clear. They were not in astronautics to solve the mysteries of the moon, they were astronauts to save America.

Nonetheless, if two-thirds of the astronauts were politicians and patriots, the remainder might still be priests of a religion not yet defined nor even discovered. One met future space men whose manner was frinedly and whose talk was small, but it was possible they had a mission. Like Armstrong or Aldrin they were far from the talk at hand. If they followed the line of a conversation, they still seemed more in communion with some silence in the unheard echoes of space.

Guided by the unlikely William Miller, minister of the Fourth Presbyterian Church of Trenton, New Jersey, a number of technicians and financial backers (all from the private sector) spent twelve years and $1.5 million to reach the moment described below. The Aereon project, in its various incarnations from model to manned vehicle, was to produce a weird combination of the airplane and the airship that would fly aerodynamically and float aerostatically and be capable of transporting enormous loads — buildings, fleets of trucks, bridges. The Aereon was bright orange, with an arching back and a deep belly. "Seen from above," writes John McPhee, "it was a delta. From the side, it looked like a fat and tremendous pumpkin seed." On this day, March 6, 1971, test pilot Jack Olcott would take the deltoid pumpkin seed out for a spin.

JOHN McPHEE

The Deltoid Pumpkin Seed

Miller had long since discovered the hole in the fence between religion and superstition. All week long, as he worried and as he watched the weather, he looked for omens. When he learned that a major reunion of American airship men would be held at Lakehurst on June 26 and 27 next, he shivered with hopeful presentiment. The aircraft registration number of Aereon 26, boldly painted on its side, was N2627. The weather forecast for Saturday, March 6th, was more than promising. All the mechanical work Olcott had asked for had been done. So word went out on Friday, through the Aereon answering service, that the test group should meet at NAFEC at five-thirty the following morning. Reaching into a pocket, Miller took out his daily appointment book. It was called the Success Agenda Seven-Star Diary, and it included a fortune message for each day. Turning the page, Miller read the message for March 6, 1971. It said, "The mocker's arrow turns back like a boomerang."

Jack Olcott and his wife, Hope, happened to be giving a dinner party March 5th, as they had on the night before the first lift-off, six months before. Olcott mixed himself an aquatini — water with an olive in it — and after dinner he passed cordials around and said to his guests, "I hope you won't consider me rude, and I hope I won't break up the conversation, but I have a very early morning appointment and I have to retire for the night." He got up at four, and took fifteen minutes to dress, choosing a blue blazer, a blue-and-gold button-down striped shirt, his royal-blue tie with fleurs-de-lis, gray flannel trousers, and a pair of defeated, broken-down loafers with flapping soles. In the blazer's lapel was a small set of wings, emblematic of his membership in the secret society of Quiet Birdmen. At four-thirty, he parked his car at Morristown Airport, and shoveled an aging snowdrift from the apron of a T-hangar. He rolled out a Beechcraft Travelair, climbed in, and took off. His route south passed above McGuire Air Force Base. The McGuire approach controller said to him, "Are you going home late or getting up early?" Olcott gave the controller a straight answer. Crossing the Pine Barrens, he ate a box breakfast that his wife had packed—crullers and coffee, meat-loaf sandwiches. At five-thirty, he raised the galactic blue lights of NAFEC.

The big hangar was crowded. The 26 was nestled like an orange-dyed egg between the wing and tail fin of a Convair 880, a four-engine commercial jet roughly the size of a 707 or a DC-8. Around the 26 was nonagon of gold nylon cord, strung among nine wooden stanchions. Linkenhoker, inside the barricade, was finishing up the preflight inspection. ("I had one major thing in my mind," he said later. "How might I feel if through some fault in the aircraft it cracked up and we lost a man? This was the foremost thought in my mind the whole time we were down there. I know one thing now: I'll never be placed in a position where I have to take complete responsibility for a man's life again. The design was good, but, nevertheless, the over-all putting together of the aircraft was mine, and that presented a hell of a feeling, I'll tell you. I thought, Here we are using an unaccepted structure and an uncertified engine, and we have low prior knowledge of the vehicle's flight characteristics. It presented a rather dismal picture in my mind. Fortunately, we were so damned busy—the buildup to the tests was so great—that I didn't have much time to think.") Everything seemed right with the aircraft and its engine—1,101 pounds minus Olcott, center of gravity 50.35 percent, examined and ready to go. Linkenhoker began to remove the gold cord. John Kukon, who had no official role to perform, had got up in the middle of the night and come to NAFEC anyway, unable to resist seeing this particular outing. Olcott, now in his test-pilot clothes, slot pockets bristling with stubby pencils, was telling Kukon stories about experimental airplanes he had known and interesting troubles they had had. There had been one in India, for example, "with a classical aileron wing-bending flutter problem" that always developed at just so many miles per hour. Olcott would accelerate the plane until he got a nice, pronounced flutter going. With a high-speed camera he would take pictures of the flutter. He also told a story about a plane that had recently crashed in a bizarre way, yielding three survivors. If these were parables, they were to Olcott himself subliminal. His manner was, as always, calm and precise. He asked Kukon what he thought about the popping in the engine. Small power plants like that were not unlike model-aircraft engines, about which there was very little that Kukon did not know. Kukon told him not to worry. The 26 had a two-cycle engine, like a chain saw or an outboard, developing a great deal of horsepower for its weight. Two-cycle engines run on combined gas and oil, lubricating themselves as they go along, and just the right amount of air has to be mixed with this fuel to produce maximum horse-power. If the mixture is too lean, horsepower declines, and—more important—the engine can develop too much heat and destroy itself. If the mixture is too rich, horse-power declines also, but the engine functions well. One sign of a rich mixture is that the engine occasionally, harmlessly, pops. Olcott was used to flying four-cycle engines, and that, of course, was another story altogether. Popping in a four-cycle engine could be a symptom of catastrophic trouble. With a two-cycle engine, though, the best ratio for the fuel–air mixture was just a little way over on the safe side of the power peak— popping now and again, like corn on a stove.

Olcott thanked Kukon and said he felt relieved of that problem. The tall, telescopic doors moved apart. The Aereon was rolled toward the breaking day. Emerging from beneath the Convair 880, the 26 seemed small to the point of absurdity, with its little chain-saw-type engine mounted above the rear like a horsefly sitting on the head of a pin. Minuscule beside the giant airplane, the Aereon was hard to imagine at full scale, but if it ever grew to its ultimate conceptual dimensions it would not be able to insert into this big hangar a great deal more than its nose, for it would be the size of the *Hindenburg* and the *Graf Zeppelin* placed together in the shape of a T, with superstruc-ture filled in to form an immense rigid delta. A couple of dozen Convair 880s could fit inside it. Linkenhoker, standing on an iron stool, primed the Aereon's engine, and tugged at a blade of the new propeller. The engine eventually coughed, ignited, and racketed against the walls of NAFEC. Olcott closed the hatch, radioed for permission to move, and routinely went up Taxiway Bravo toward the head of the runway.

For stopwatch timing and for photography, Miller, Putman, and others were deliv-ered by station wagon to various points on the airfield. Fire and crash vehicles were operating on both sides of the runway this time—yellow lights, red lights flashing everywhere. The morning was pale blue, clear, and fine. The sun was above the hori-zon, its light streaming to the west. "This is a day the Lord has made," Miller said. "Let us rejoice and be glad in it. It's just ideal. No wind. Dry. Clear. This day is a gift." Olcott was facing west, at the head of the runway. Cleared, he accelerated, rotated, and lifted into the air. He climbed to forty feet and leveled off. He tried a coordinated bank and gentle turn to the left. It went well. He did the same to the right. The 26 responded as he had expected it would. It did not seem to have a tendency to roll excessively. The roll damping was light. He got a promising sense of the roll-control effectiveness of the vehicle. Reducing power over the seven-thousand-foot markers, he descended, landed, and taxied to the turnaround block at the runway's western end. The brief flight had been Olcott's warmup. He now felt that he had the vehicle all around him. NAFEC had asked him to take off to the east if he ever intended to leave the airspace of the runway, because the terrain at the eastern end of the NAFEC reservation had a dirt road winding through it and was particularly accessible to fire trucks and crash vehi-cles. Olcott was now facing east. He called the tower and said this was Aereon 2627 requesting permission to lift off and make a circuit of the field. Two miles of broad runway reached out before him. The parallel taxiway, where the fire trucks would race him, was to his right. To the right of the taxiway and a little more than halfway down was the great dark block of the NAFEC hangar, its near wall lined with tiers of offices behind shining plate-glass windows that reflected the low rays of the sun. The fire trucks and other cars were lined up and ready. Olcott showed them a raised thumb, moved the engine up to four thousand revolutions per minute, and left the head of the runway. He watched the black tape marks on his windshield, rotated, established his angle of attack, and went into the air. He climbed to forty feet. This time, however, he did not level off. Still in a position to abort the flight with ease, still in an environment he had been in many times before, he now had to decide whether the rate of climb was sufficient to warrant an advance to where he had not been. He had hoped for a rate of climb of 200, or even 250, feet per minute, but the engine was running flat out and he was getting 150. He figured that 100 feet per minute, or less, would be so marginal that he would have to go down. This was, for sure, the inverse frontier — an exploration of the lower, most economical limits of aerodynamic possibility. Some sommercial jets climb six thousand feet per minute. He watched the rate-of-climb indicator. It was holding at 150 positive rate of climb — positive enough for him to decide to stay with it. He put in a little rudder and made a slight right turn. Moving obliquely, he would add something to the time when he would be near enough to the runway to get to it if the engine failed. He was flying directly toward the NAFEC hangar, however, and NAFEC sternly told him to head somewhere else at once. He was about eighty feet in the air. If his engine failed, he could not have hit the NAFEC building even if he tried. The building was half a mile away. The 26 had a glide ratio of about five to one. From that altitude, the 26 could not have glided more than four hundred feet before scraping the ground. Moreover, the 26 was so light that if it had hit the building head on it might have had difficulty breaking the glass. Olcott corrected his turn, though, and continued to climb slowly to the east. It was like driving a station wagon stuffed with cordwood up the side of a mountain in first gear. He was getting there. He knew he would make it over the hill. Meanwhile, there was nothing to do but be patient. He reached 100 feet, 150 feet, 200 feet, all the while reminding himself: Do not change anything. Stay at this airspeed. Hold the controls with constant pressure. Let the vehicle do the work. These tests are important, they must be concluded. You knew all along that the vehicle was never going to behave like a homesick angel. It just wasn't going to climb like that. Within the margins of our considerations was a poor rate of climb.

The 26 was almost over the end of the runway, and was 250 feet in the air. Seen from the ground and from a mile behind, it appeared to be a small black diamond moving into the sun. "Fantastic!" John Kukon said. "It's got a lot higher nose attitude than I expected, but if that's the way it is, so be it."

Olcott was now about to try the first significant turn the 26 had ever made. He could not with certainty predict what would happen. He did not have the altitude he had planned for, and he had to ask himself a lot of questions. He had to keep looking for and selecting places where he might set the Aereon down if the engine stopped, or if much of anything else went wrong. You can't wait until the engine quits to decide how to handle the situation. It's too late then. You have to know what you're going to do before you have to do it. So you are continually saying to yourself: What will I do if this happens? What will I do if that happens? If the engine quits now, I'll put the stick forward to make sure I'm going downhill, like the boy on a bicycle who doesn't want to pedal anymore. He's got to be pointed downhil or he'll topple over. Get the nose down. Establish the glide. Keep the airspeed the same, so you have control. Then go into one of those preselected landing spots. The engine is the primary consideration. Stability is the secondary consideration. The Aereon is not a broom balancing on the palm of your hand. It is a stable vehicle. Nevertheless, you do not yet know to what extent it is stable. As long as you don't disturb anything—as long as you move into any control input very slowly and smoothly—the chances are that you'll never upset the dynamics of the vehicle so drastically that you cannot cope with it. I am 250 feet over the end of the runway. If the engine fails here, there is no way I can turn around and get back into the runway. If I were to try, I'd probably lose control of the aircraft. So what do I do? Where would I go? The dirt road. It is sort of a hard dirt road. I believe I could get in there. Maybe damage the nose gear but not do too much harm. I could negotiate that landing.

He went into the turn. He made it shallow, because he had never been in one before. His mind raced with the conditions and problems of the turn, addressing himself, addressing the aircraft. Let's take it nice and easy. Let's not depart too much from what we've done before. Here we go. This is the first time we've really got a sustained angle of bank. Really a turn. We know from the computer simulations that if the angle of bank gets a little too high, and the rudders are not coordinated just right, the vehicle will want to continue to the left and will be difficult to control. That would be disconcerting at low altitude. There's a straightforward way out, with use of rudders and manipulation of the stick.

He had taped one end of a bit of black yarn to the outside of the cockpit canopy, and now he watched it closely. Air should always be flowing straight back, no matter what maneuver the aircraft might be making. If the yarn were to move sidewise, the 26 would be going into a yaw. The yarn was straight. The sideslip angle was zero—just what it was supposed to be.

The 26, continuing to climb, had turned through an arc of ninety degrees and was heading north. Olcott no longer needed the dirt road. If trouble developed now, he could probably get around to the runway, heading west, if he had to. It was like trying to cross a stream from one bare rock to the next bare rock, trying not to fall in. Meanwhile, in addition to and above all else, he was supposed to be collecting test data. What is the rate of climb now? What is the indicated airspeed? What is the angle of attack? What is the control-position transducer saying? How am I doing? How am I doing relative to what I want to be doing? How much will this turn hurt the rate of climb?

The 26 completed its wide arc to 180 degrees and was headed west, parallel to, but considerably north of, the runway. The rate of climb had remained steady. The ship was 400 feet up now, and it continued to rise until Olcott leveled off, as he had planned to, at 500 feet. Data now flowed from the instruments. The maximum speed, full throttle, was 64 knots—a little better than Olcott had expected. He planned his route over the western end of the reservation, telling himself not to fly over the houses there, because that was not good professional technique in an aircraft that had a limited flight history and a configuration that had never flown before. I'll just have to go into the shrubbery if anything happens here, he told himself but he swung into a perfect 180 degree turn and was now pointed again into the sun. He was 500 feet over the broad white stripes from which he had begun his takeoff. He had completed a circuit of the field.

"Wow!" Miller said, shooting straight up with his Nikon Super 8. "This is fantastic!" The fire trucks stopped running around. All the ground vehicles stopped. Everyone watched the sky.

Olcott now had the Atlantic Ocean spread before him, wide marshes and bays, the skyline of Atlantic City to his right, Absecon Bay straight ahead, and to the left the Brigantine National Wildlife Refuge. Almost below him was the Garden State Parkway, a superior alternative to the dirt road as an emergency landing strip. Northbound or southbound, the 26 could blend right in with the cars there, if necessary, at an identical speed. Olcott found the scope of his view extraordinary, because there were no wings around him to impede it.

Olcott again circled the field, this time reducing his airspeed to 59 knots to see how the 26 would handle there. Then he went around again, at 52 knots, and again, at 50. Each circuit was about eight miles.

"It's slowly sinking in," Miller said. "He's not going to come down."

In subsequent days, Olcott would fly the 26 right out to the end of its engine time. It would be tracked by NAFEC's theodolite, yielding, for $200 an hour, precise airspeed data. Olcott would do Dutch rolls and steady sideslips, kicking out hard with his rudders. He would do aileron rolls to the right, aileron rolls to the left, rudder kicks right, rudder kicks left, as if he were practicing swimming. He would climb, slow down, dive, speed up—a fundamental longitudinal mode, the phugoid motion. Investigating the phugoid, he would go into a steady sideslip and then "put in a doublet—just to get the thing excited."

"That's how we lost one of the Aereon 4s," Linkenhoker would say, biting a toothpick, watching from the ground, and then, perhaps because he was unable just to stay there and watch, Linkenhoker would jump into a Piper Cherokee and chase the 26 into the sky. I went with him. The 26 seemed to float beside us, over the field, pinewoods, the parkway—with tidal estuaries, salt marshes, and the sea beyond. Shafts of sunlight sprayed down from behind clouds in which the sun kept appearing as a silver disc, and, moving in and out of these palisades of light, the 26 went into smooth roll angles and controlled yaws — part airplane, part airship, floating, flying, settling in to landings light and slow. "Aereon is great," said NAFEC's chief executive officer. "Just look at it and you can see the potential. What made New York great? What made Chicago great? The carrying of freight." One could almost see New Yorks and Chicagos springing up under the slow-moving shadow of the Aereon as it flew. A subtler and perhaps more durable endorsement had come from NAFEC beforehand, however, on the day of the first circuit of the field. Flying on and on — the first circuiting flight lasted more than half an hour — Olcott looked up at one point to see a Starlifter approaching the field. The two aircraft — one weighing 1,100 pounds, the other weighing 70 tons — were more or less on a collision course. "Tell them to give me plenty of room," Olcott said to the tower. "I cannot tolerate their wake." The tower told the Starlifter to turn right, go south, and keep on going south indefinitely. "The traffic on your left," the tower explained, "is an aerobody — a wingless vehicle — proceeding northwest."

This story from *Tales of the South Pacific* (1947) provides neat counterpoint to the earlier excerpt from Joseph Heller's *Catch-22*. Different theater of operations, but same war, same subject, same sin of overconfidence.

JAMES MICHENER

The Milk Run

It must make somebody feel good. I guess that's why they do it. The speaker was Lieutenant Bus Adams, SBD pilot. He was nursing a bottle of whisky in the Hotel De Gink on Guadal. He was sitting on an improvised chair and had his feet cocked up on a coconut stump the pilots used for a footrest. He was handsome, blond, cocky. He came from nowhere in particular and wasn't sure where he would settle when the war was over. He was just another hot pilot shooting off between missions.

But why they do it—Bus went on—I don't rightfully know. I once figured it out this way: Say tomorrow we start to work over a new island, well, like Kuralei. Some day we will. On the first mission long-range bombers go over. Sixty-seven Japs come up to meet you. You lose four, maybe five bombers. Everybody is damn gloomy. I can tell you. But you also knock down some Nips.

Four days later you send over your next bombers. Again you take a pasting. "The suicide run!" the pilots call it. It's sure death! But you keep on knocking down Nips. Down they go, burning like the Fourth of July. And all this time you're pocking up their strips, plenty.

Finally the day comes when you send over twenty-seven bombers and they all come back. Four Zekes rise to get at you, but they are shot to hell. You bomb the strip and the installations until you are dizzy from flying in circles over the place. The next eight missions are without incident. You just plow in, drop your stuff, and sail on home.

Right then somebody names that mission, "The Milk Run!" And everybody feels pretty good about it. They even tell you about your assignments in an offhand manner: "Eighteen or twenty of you go over tomorrow and pepper Kuralei." They don't even brief you on it, and before long there's a gang around takeoff time wanting to know if they can sort of hitchhike a ride. They'd like to see Kuralei get it. So first thing you know, it's a real milk run, and you're in the tourist business!

Of course, I don't know who ever thought up that name for such missions. The Milk Run? Well, maybe it is like a milk run. For example, you fill up a milk truck with TNT and some special detonating caps that go off if anybody sneezes real loud. You tank up the truck with 120 octane gasoline that burns Pouf! Then instead of a steering wheel, you have three wheels, one for going sideways and one for up and down. You carry eight tons of your special milk when you know you should carry only five. At intersections other milk trucks like yours barge out at you, and you've got to watch them every minute. When you try to deliever this precious milk, little kids are all around you with .22's, popping at you. If one of the slugs gets you, bang! There you go, milk and all! And if you add to that the fact that you aren't really driving over land at all, but over the ocean, where if the slightest thing goes wrong, you take a drink. . . . Well, maybe that's a milk run, but if it is, cows are sure raising hell these days!

Now get this right. I'm not bitching. Not at all. I'm damned glad to be the guy that draws the milk runs. Because in comparison with a real mission, jaunts like that really *are* milk runs. But if you get bumped off on one of them, why you're just as dead as if you were over Tokyo in a kite. It wasn't no milk run for you. Not that day.

You take my trip up to Munda two days ago. Now there was a real milk run. Our boys had worked that strip over until it looked like a guy with chicken pox, beriberi, and the galloping jumps. Sixteen SBD's went up to hammer it again. Guess we must be about to land somewhere near there. Four of us stopped off to work over the Jap guns at Segi Point. We strafed them plenty. Then we went on to Munda.

Brother, it was a far cry from the old days. This wasn't The Slot any more. Remember when you used to bomb Kieta or Kahili or Vella or Munda? Opposition all the way. Japs coming at you from every angle. Three hundred miles of hell, with ugly islands on every side and Japs on every island. When I first went up there it was the toughest water fighting in the world, bar none. You were lucky to limp home.

Two days ago it was like a pleasure trip. I never saw the water so beautiful. Santa Ysabel looked like a summer resort somewhere off Maine. In the distance you could see Choiseul and right ahead was New Georgia. Everything was blue and green and there weren't too many white ack-ack puffs. I tell you, I could make that trip every day with pleasure.

Segi Point was something to see. The Nips had a few antiaircraft there, but we came in low, zoomed up over the hills, peppered the devil out of them. Do you know Segi Passage? It's something to remember. A narrow passage with maybe four hundred small pinpoint islands in it. It's the only place out here I know that looks like the South

Pacific. Watch! When we take Segi, I'm putting in for duty there. It's going to be cool there, and it looks like they got fruit around, too.

Well, after we dusted Segi off we flew low across New Georgia. Natives, and I guess some Jap spotters, watched us roar by. We were about fifty feet off the trees, and we rose and fell with the contours of the land. We broke radio silence, because the Japs knew we were coming. The other twelve were already over target. One buddy called out to me and showed me the waterfall on the north side of the island. It looked cool in the early morning sunlight. Soon we were over Munda. The milk run was half over.

I guess you heard what happened next. I was the unlucky guy. One lousy Jap hit all day, on that whole strike, and it had to be me that got it. It ripped through the rear gunner's seat and killed Louie on the spot. Never knew what hit him. I had only eighty feet elevation at the time, but kept her nose straight on. Glided into the water between Wanawana and Munda. The plane sank, of course, in about fifteen seconds. All shot to hell. Never even got a life raft out.

So there I was, at seven-thirty in the morning, with no raft, no nothing but a life belt, down in the middle of a Japanese channel with shore installations all around me. A couple of guys later on figured that eight thousand Japs must have been within ten miles of me, and I guess that not less than three thousand of them could see me. I was sure a dead duck.

My buddies saw me go in, and they set up a traffic circle around me. One Jap barge tried to come out for me, but you know Eddie Callstrom? My God! He shot that barge up until it splintered so high that even I could see it burst into pieces. My gang was over me for an hour and a half. By this time a radio message had gone back and about twenty New Zealanders in P-40's took over. I could see them coming a long way off. At first I thought they might be Jap planes. I never was too good at recognition.

Well, these New Zealanders are wild men. Holy hell! What they did! They would weave back and forth over me for a little while, then somebody would see something on Rendova or Kolombangara. Zoom! Off he would go like a madman, and pretty soon you'd see smoke going up. And if they didn't see anything that looked like a good target, they would leave the circle every few minutes anyway and raise hell among the coconut trees near Munda, just on chance there might be some Japs there. One group of Japs managed to swing a shore battery around to where they could pepper me. They sent out about seven fragmentation shells, and scared me half to death. I had to stay there in the water and take it.

That was the Jap's mistake. They undoubtedly planned to get my range and put me down, but on the first shot the New Zealanders went crazy. You would have thought I was a $90 million battleship they were out to protect. They peeled off and dove that installation until even the trees around it fell down. They must have made the coral hot. Salt water had almost blinded me, but I saw one P-40 burst into flame and plunge deeply into the water off Rendova. No more Jap shore batteries opened up on me that morning.

Even so, I was having a pretty tough time. Currents kept shoving me on toward Munda. Japs were hidden there with rifles, and kept popping at me. I did my damnedest, but slowly I kept getting closer. I don't know, but I guess I swam twenty miles that day, all in the same place. Sometimes I would be so tired I'd just to have stop, but whenever I did, bingo! There I was, headed for the shore and the Japs. I must say, though, that Jap rifles are a damned fine spur to a man's ambitions.

When the New Zealanders saw my plight, they dove for that shoreline like the hounds of hell. They chopped it up plenty. Jap shots kept coming after they left, but lots fewer than before.

I understand that it was about this time that the New Zealanders' radio message reached Admiral Kester. He is supposed to have studied the map a minute and then said, "Get that pilot out there. Use anything you need. We'll send a destroyer in, if necessary. But get him out. Our pilots are not expendable."

Of course, I didn't know about it then, that was mighty fine doctrine. So far as I was concerned. And you know? When I watched those Marine F4U's coming in to take over the circle, I kind of thought maybe something like that was in the wind at headquarters. The New Zealanders pulled out. Before they went, each one in turn buzzed me. Scared me half to death! Then they zoomed Munda once more, shot it up some, and shoved off home.

The first thing the F4U's did was drop me a life raft. The first attempt was too far to leeward, and it drifted toward the shore. An energetic Jap tried to retrieve it, but one of our planes cut him to pieces. The next raft landed above me, and drifted toward me. Gosh, they're remarkable things. I pulled it out of the bag, pumped the handle of the CO_2 container, and the lovely yellow devil puffed right out.

But my troubles were only starting. The wind and currents shoved that raft toward the shore, but fast. I did everything I could to hold it back, and paddled until I could hardly raise my right arm. Then some F4U pilot with an IQ of about 420—boy, how I would like to meet that guy—dropped me his parachute. It was his only parachute and from then on he was upstairs on his own. But it made me a swell sea anchor. Drifting far behind in the water, it slowed me down. That Marine was a plenty smart cookie.

It was now about noon, and even though I was plenty scared, I was hungry. I broke out some emergency rations from the raft and had a pretty fine meal. The Jap snipers were falling short, but a long-range mortar started to get close. It fired about twenty shots. I didn't care. I had a full belly and a bunch of F4U's upstairs. Oh, those lovely planes! They went after that mortar like a bunch of bumblebees after a tramp. There was a couple of loud garummmphs, and we had no more trouble with that mortar. It must have been infuriating to the Japs to see me out there.

I judge it was about 1400 when thirty new F4U's took over. I wondered why they sent so many. This gang made even the New Zealanders look cautious. They just shot up everything that moved or looked as if it might once have wanted to move. Then I saw why.

A huge PBY, painted black, came gracefully up The Slot. I learned later that it was Squadron Leader Grant of the RNZAF detachment at Halavo. He had told headquarters that he'd land the Cat anywhere there was water. By damn, he did, too. He reconnoitered the bay twice, saw he would have to make his run right over Munda airfield, relayed that information to the F4U's and started down. His course took him over the heart of the Jap installations. He was low and big and a sure target. But he kept coming in. Before him, above him, and behind him a merciless swarm of thirty F4U's blazed away. Like tiny, cruel insects protecting a lumbering butterfly, the F4U's scoured the earth.

Beautifully the PBY landed. The F4U's probed the shoreline. Grant taxied his huge plane toward my small raft. The F4U's zoomed overhead at impossibly low altitudes. The PBY came alongside. The F4U's protected us. I climbed aboard and set the raft loose. Quickly the turret top was closed. The New Zealand gunner swung his agile gun about. There were quiet congratulations.

The next moment hell broke loose! From the shore one canny Jap let go with the gun he had been saving all day for such a moment. There was a ripping sound, and the port wing of the PBY was gone! The Jap had time to fire three more shells before the F4U's reduced him and his gun to rubble. The first two Jap shells missed, but the last one blew off the tail assembly. We were sinking.

Rapidly we threw out the rafts and as much gear as we could. I thought to save six parachutes, and soon nine of us were in Munda harbor, setting our sea anchors and looking mighty damned glum. Squadron Leader Grant was particularly doused by the affair. "Second PBY I've lost since I've been out here," he said mournfully.

Now a circle of Navy F6F's took over. I thought they were more conservative than the New Zealanders and the last Marine gang. That was until a Jap battery threw a couple of close ones. I had never seen an F6F in action before. Five of them hit that battery like Jack Dempsey hitting Willard. The New Zealanders, who had not seen the F6F's either, were amazed. It looked more like a medium bomber than a fighter. Extreme though our predicament was, I remember that we carefully appraised the new F6F.

"The Japs won't be able to stop that one!" an officer said. "It's got too much."

"You mean they can fly that big fighter off a ship?" another inquired.

"They sure don't let the yellow bastards get many shots in, do they?"

We were glad of that. Unless the Jap hit us on first shot, he was done. He didn't get a second chance. We were therefore dismayed when half of the F6F's pulled away toward Rendova. We didn't see them anymore. An hour later, however, we saw thirty new F4U's lollygagging through the sky Rendova way. Four sped on ahead to relieve the fine, battle-proven F6F's who headed down The Slot. We wondered what was up.

And then we saw! From some secret nest in Rendova, the F4U's were bringing out two PT boats! They were going to come right into Munda harbor, and to hell with the Japs! Above them the lazy Marines darted and bobbed, like dolphins in an aerial ocean.

You know the rest. It was Lieutenant Commander Charlesworth and his PT's. Used to be on Tulagi. They hang out somewhere in the Russells now. Something big was on, and they had sneaked up to Rendova, specially for an attack somewhere. But Kester shouted, "To hell with the attack. We've gone this far. Get that pilot out of there." He said they'd have to figure out some other move for the big attack they had cooking. Maybe use destroyers instead of PT's.

I can't tell you much more. A couple of savvy Japs were waiting with field pieces, just like the earlier one. But they didn't get hits. My God, did the Marines in the F4U's crucify those Japs! That was the last thing I saw before the PT's pulled me aboard. Twelve F4U's diving at one hillside.

Pass me that bottle, Tony. Well, as you know, we figured it all out last night. We lost a P-40 and a PBY. We broke up Admiral Kester's plan for the PT boats. We wasted the flying time of P-40's, F4U's, and F6F's like it was dirt. We figured the entire mission cost not less than $600,000. Just to save one guy in the water off Munda. I wonder what the Japs left to rot on Munda thought of that? $600,000 for one pilot. — Bus Adams took a healthy swig of whisky. He lolled back in the tail-killing chair of the Hotel De Gink. — But it's sure worth every cent of the money. If you happen to be that pilot.

Despite the sensational sinkings of surplus battleships which he arranged from 1921 to 1923, Brigadier General William "Billy" Mitchell could not convince his superiors that dominance in the air was the key to winning any war in the future. His continuing agitation on behalf of his cause included harsh words that led to his court-martial in 1925. Despite his forced return to civilian life, Mitchell continued to promote air power in such books as *Skyways* (1930), source of the passage below. Mitchell died in 1936, but when Americans raided Tokyo in April 1942, they flew in B-25 "Mitchell" bombers named in his honor.

BILLY MITCHELL

The Kind of Men to Make Into Pilots

Most healthy young men or women from sixteen to forty years of age can be taught to fly an ordinary airplane. A great majority of these may become very good pilots for transport- or passenger-carrying machines in time of peace; but the requirements for a military aviator call for more concentrated physical and mental ability in the individual than has ever been necessary in any calling heretofore.

The military flyer must move at tremendous speed through the air. Not only must he fly his own airplane, but he must see all the other airplanes around him, friendly as well as hostile ones; he must observe the ground, be ready to attack or defend himself at any instant, be ready to land on any sort of surface, mountains, forests, plowed ground. He may have to fly for hundreds or thousands of miles to his destination and there deliver his cargo or bombs, or report what he sees, or make his attack either against hostile air forces, vessels on the sea or vital centers. It is a supreme test of character and courage. Often no one is watching. If he shows the white feather or is inclined to "beat" what he is instructed to do, often no one can be the wiser. When he returns he may tell a fantastic tale about things that never occurred without much danger of being checked up.

I have seen this done in many instances. On one notable occasion, the pilot shot all his ammunition away into the empty air and returned to report that he was out of ammunition. This resulted in the death of one of our greatest aviators because he was left without assistance and had to close with the enemy alone. In other cases, I have known men who were excellent pilots behind the lines but the instant they came into the presence of the enemy, either their machine guns jammed or they would have engine trouble or something that required their leaving that vicinity immediately. They would land with a plausible story which was impossible to disprove. Other men were perfectly free in their admission that when they came into the presence of the enemy, they were absolutely unable to control themselves. These men were brave, resourceful and patriotic but they did not have the moral qualities required for that kind of work.

With an army on the ground, a person is shoulder to shoulder with another. If he is wounded, he can lie down on the ground and be aided by his comrades. If he is on the sea, he is in a ship close to his companions. When in the air, however, he is away from these, off by himself thousands of feet up. If he is not successful, he goes down in flames. These things are not so pleasant to talk about, but they are the things that affect a man's mind.

The American makes a particularly good pilot because he comes of sturdy stock, with the pioneering instinct of doing something new still in him. He is brought up in our schools and universities to play games that require a good deal of courage, particularly American football. The other games, such as baseball, basketball, hockey, and tennis, not only teach him individual initiative and self-reliance but to rely on his comrades, to work together in a team and to acquire the kind of discipline necessary in the air. This is "thinking discipline" in which the individual estimates exactly what his orders mean and then carries them out by using his head in accordance with the circumstances that exist.

The old discipline, as conceived and carried out by armies and navies throughout the centuries, consists in the unhesitating obedience by a subordinate to the orders of his superior. In that case the subordinate is not supposed to think too much. If he does, he may "spill the beans" and tie up the whole operation. This kind of discipline has been designed for the average man. It is only within the last generation that most of the men composing armies could read or write. The development of their mentality on the average was of a comparatively low order. With the aviator, however, the keenest, best-educated, most advanced type of man has to be selected. These are some of the reasons why an officer raised in any army or navy atmosphere is totally incapable of understanding an airplane pilot's mentality or the way to handle or develop him.

The development of aviation and its undoubted assumption of the principal means of national defense in which a comparatively small number of individuals will do the work, will result in a condition similar to that during the Middle Ages when the fighting was done by a few knights in armor while the rest of the people supported them.

Today the nation that has suitable men to make into aviators and the industrial background and resources to create airplanes and equipment will certainly dominate the future world. Like everything else, the execution will depend on the individual.

Physically, the doctors tell us a great deal about how the man should be constituted. A great many of their deductions are correct. They say he must be sound in the eyes, in the heart and lungs. His inner ear must be capable of determining balance. He must be free from disease, must have good hands and feet, legs, arms, and body, and probably hair on his head. But some of the greatest aviators we have ever had have been afflicted with tuberculosis, have had only one eye, have had a club foot, have had bad hearts and been otherwise impaired.

We hear a great deal about the age at which people have to begin or stop flying. This also is relative. I have seen men begin to fly in their late thirties who became extraordinary pilots. Naturally, when a man gets over forty his eyes begin to lose the power of accommodation, but this only means that they are unable to focus quickly on near objects at ordinary reading distances, that is, twenty inches from the eyes. A man who has sound eyes at any age can see perfectly well for flying. Of course if his eyes have astigmatism, he is always burdened with them. I know some men whose vision was astigmatic, who memorized the eye tests and "got by" with them, and made excellent pilots. On the other hand, there were other men who did the same thing and it was directly responsible for their death. My own brother met his death in an airplane for that reason.

It was interesting to see how pilots were trained by the different countries during the war to get the most out of them. One country did not teach its pilots what the dangers were. It took them when they were very young and consequently inexperienced, and instilled into them the idea of closing with the enemy and destroying him, irrespective of loss. Their theory was that men taught in this way, no matter if they did incur greater losses, would destroy a great many of the enemy. If their side had a greater number of pilots, they would win in the end; but it did not work out that way.

Another nation taught its pilots to be too cautious, and enlarged too much on the dangers of air work, consequently they did not do as much as they should. I have always tried to teach our pilots every side of the question and then to have them push forward to the accomplishment of their tasks with a thorough knowledge of everyting that lay in front of them.

The LZ-129, better known as the *Hindenburg*, was built as the ultimate airship, an improvement upon even the spectacularly successful *Graf Zeppelin*. It was a beautiful and luxurious craft, the culmination of Count Ferdinand von Zeppelin's dream. In 1936 the *Hindenburg* made ten successful round trips to the United States; despite the American ban on exporting helium, a glorious future for the giant airships seemed assured. Then, in thirty-two seconds over Lakehurst, New Jersey, on May 6, 1937, the age of the airship came to a horrifying end. Reporting on the scene was Herb Morrison of radio station WLS. These are his words.

HERB MORRISON

"Oh, the Humanity!"

Here it comes, ladies and gentlemen, and what a sight it is, a thrilling one, a marvelous sight. It is coming down out of the sky pointed toward us, and toward the mooring mast. The mighty diesel motors roar, the propellers biting into the air and throwing it back into galelike whirlpools. Now and then the propellers are caught in the rays of the sun, their highly polished surfaces reflect. . . .

No one wonders that this great floating palace can travel through the air at such a speed with these powerful motors behind it. The sun is striking the windows of the observation deck on the eastward side and sparkling like glittering jewels on the background of black velvet. . . .

taken hold of by a number of men on the field. It is starting to rain again; the rain had slacked up a little bit. The back motors of the ship are holding it just enough to keep it

—IT'S BURST INTO FLAME!

Get out of the way! Get this, Charlie, get this, Charlie . . . get out of the way *please!* Oh my, this is terrible, oh my, get out of the way, *please!* It is burning, bursting into flames and is falling on the mooring mast and all the folks we . . . this is one of the worst catastrophes in the world! Oh, the flames are four or five hundred feet into the sky, it's a terrific sight, ladies and gentlemen. It is in smoke and flames now. Oh, the humanity and all the passengers!

I can't talk, ladies and gentlemen. Honest, it is a mass of smoking wreckage. Lady, I am sorry. Honestly, I can hardly—I am going to step inside where I can't see it. Charlie, that is terrible!

Listen, folks, I am going to have to stop for a minute, for it's the worst thing I have ever witnessed.

After constructing the great labyrinth which housed the Minotaur on the island of Crete, the inventor Daedalus and his son, Icarus, wished to return to their native Greece; King Minos would not let them leave. The rest is legend, here told by the Roman poet Ovid in his *Metamorphoses* of A.D. 8. The translation is by Samuel Croxall.

OVID

The Fall of Icarus

In tedious exile
   now too long detain'd,
Daedalus languish'd
   for his native land;
The sea foreclosed his flight,
   yet thus he said:
"Though earth and water
   in subjection laid,
O cruel Minos, thy dominion be,
We'll go through air;
   for sure the air is free."
Then to new arts
   his cunning thought applies,
And to improve
   the work of nature tries.
A row of quills
   in gradual order placed,
Rise by degrees
   in length from first to last. . . .

Along the middle
   runs a twine of flax,
The bottom stems
   are join'd by pliant wax:
Thus, well compact,
   a hollow bending brings
The fine composure into real wings.
His boy, young Icarus,
   that near him stood,
Unthinking of his fate,
   with smiles pursued
The floating feathers. . . .
Or with the wax
   impertinently play'd,
And, with his childish tricks,
   the great design delay'd.
The final master-stroke
   at last imposed,
And now the neat machine
   completely closed;
Fitting his pinions on,
   a flight he tries,
And hung, self-balanced,
   in the beaten skies.
Then thus instructs his child:
   "My boy, take care
To wing your course
   along the middle air:
If low, the surges
   wet your flagging plumes;
If high, the sun
   the melting wax consumes. . . .
But follow me: let me
   before you lay
Rules for the flight,
   and mark the pathless way."
Then, teaching,
   with a fond concern, his son,
He took the untried wings
   and fix'd them on. . . .
When now the boy,
   whose childish thoughts aspire

To loftier aims,
   and make him ramble higher,
Grown wild and wanton,
   more imbolden'd, flies
Far from his guide,
   and soars among the skies.
The softening wax,
   that felt a nearer sun,
Dissolved apace,
   and soon began to run;
The youth in vain
   his melting pinions shakes,
His feathers gone,
   no longer air he takes;
O! father, father!
   as he strove to cry,
Down to the sea
   he tumbled from on high,
And found his fate;
   yet still subsists by fame
Among those waters
   that retain his name.

This discussion of Leonardo's approach to the problem of flight is a model of historical scholarship—clear, concise, and thought-provoking. It is taken from the chapter "Leonardo, Civil Engineer" in William Barclay Parsons's masterwork of 1939, *Engineers and Engineering in the Renaissance*.

WILLIAM BARCLAY PARSONS

Leonardo and the Mysteries of Flight

From early antiquity man has desired to use the air as a medium for travel, but until Leonardo no one had seriously investigated the problem or tried to reach a solution. The designing of an airplane afforded a singular opportunity for Leonardo to apply his deduced principles of mechanics—resolution of forces, stresses, and strength of materials—for in a machine that was to support itself in the air, practice had to conform strictly to theoretical requirements in order to keep the weight to the minimum consistent with efficiency.

Leonardo began, as would be expected of one possessing so logical a mind, by studying birds and other winged animals in flight. Mechanically and anatomically he determined the structure of their wings and how the various parts functioned. He avoided one pitfall into which so many would-be investigators fell: he did not think that birds sustained themselves by the beating of their wings, because he noted that birds could rest in the air and traverse quite long distances without any wing movement whatever.

He studied the action of birds conscientiously and minutely, not with any preconceived notion, but with scientific honesty, to ascertain facts and through them to learn the theory of flying. He noted their takeoff and alighting, their flight and soaring, their spiral course and their reaction to wind. From these observations he deduced some of the principles, the modern application of which has made aviation possible.

1. Flying is due to air resistance, and the air exercises as much force on an object as an object does on the air.

2. There are two centers, that of gravity and that of pressure, and their relative position to each other must be adjusted to secure equilibrium.

3. There are air currents too fine for human detection, whose existence and movement birds are able to observe and utilize by soaring and drifting with wings outstretched and motionless.

4. Birds' wings have been made by nature convex above and concave below so as to facilitate upward flight.

5. Hills and other surface variations produce "air eddies and whirlwinds," or what aviators now call pockets.

So long as Leonardo confined himself to studying birds in flight and to drawing conclusions from what he observed, he was on firm ground. When, however, he attempted to reverse his deductions and to put them in practice, his troubles began. What he lacked, and the lack was fatal to success, was a motor by which he could drive himself with stationary wings and so make use of the first of his conclusions—support is due to air resistance. But the world waited four hundred more years for the motive force that Leonardo sought.

He was cognizant of the difference between machines lighter and heavier than air, because Vasari records that Leonardo made balloons of thin wax which rose when filled with warm air. But it was the type which, like a bird or thing of life, could rise in spite of its weight and travel in any direction, that appealed to his imagination. As he put it, "Man with his large folded wings can raise himself above the air and subjugate it to him by exercising force on the resisting elements."

Lacking an engine, he had to obtain power through muscular effort working artificial wings. In an early design he took the wings of a bat for his model, wings that were smooth, light, and wide, because he pointed out that a smooth, unperforated flying surface is more suitable to mechanical flying than the wings of a feathered bird. The wing that he had in mind had a frame of pine, "which is light," covered next to the main member with fustian to which feathers had been glued to keep air from passing through it, and beyond that a light taffeta coated with starch. Figure 1 shows the wing as Leonardo sketched it, with the frame and the two coatings marked a, b, and c respectively. The letters are reversed according to Leonardo's custom.

To get a measure of the required bearing surface, in order to proportion the area of his wings to support a man, Leonardo suggested in his notes a number of experiments on large models with weights as great as 200 libbre [145 pounds].

To flap the wings, Leonardo again reverted to the bird as his model and accordingly designed a jointed frame closely resembling the main bones of a bird's wing. Figure 2 is his design of a four-jointed rod through which he hoped to transmit desired movement, to be obtained by two cords, one for raising and the other for lowering. A diagram of the cords is shown in the lower part of the illustration.

It is obvious that if motion in one direction was given by the pull on a cord and at the same time a spring was compressed, the spring would, on release of the cord, automatically give opposite motion, but Leonardo was skeptical of the efficiency of springs, preferring the direct and positive action of the two cords to obtain the up and down movement. He, therefore, wrote "No spring used." However, in a subsequent note he said that if a spring were used, the upper cord might be omitted. Figure 3 is his diagram of the joint hinge, the joints being lashed with strips of leather.

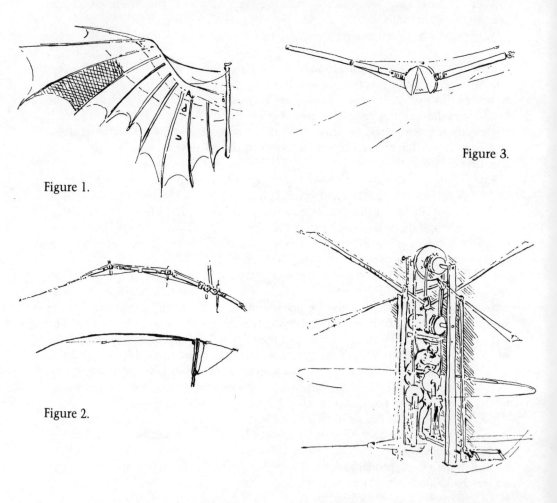

Figure 1.

Figure 2.

Figure 3.

Figure 4.

Designing the wings and their operating devices was the easiest of Leonardo's labors. The real difficulty began when he undertook to find a way to apply power through some form of human energy, nothing else being available or even within the range of his vast intellect to invent.

Many of his sketches and computations clearly indicate that he did not underestimate the amount of power needed to sustain a man and a heavier-than-air machine in flight. To obtain the maximum of human effort, he designed one machine (Fig. 4) in which the operator stood with part of the frame resting on his head, so that he exerted his strength by pushing upward at the same time that he turned a hand crank. Leonardo calculated that the push was equal in effect to 200 libbre, or as much as a man could exert through the cranks. The wings, of which there were four, were in the form of two crosses and worked alternately "like the trotting gait of a horse." He recorded that this arrangement was "better than any other."

Figures 5A and 5B show two variants of a frame or fuselage in which the aviator lay horizontally. This type he developed more in detail than the one in which the operator stood erect, and it appears to be the one that he finally accepted as preferable. In figure 5A the man stretched out on a board with the two loops passing over his body, and he exerted his power by drawing up his legs and kicking back, thereby producing tension on the wing cords. The hands were left free to manipulate the elevating and steering devices. The legs and feet could be made to work in unison and singly and so to approximate the flight of the "hawk and other birds."

In Figure 5B there was no board, and the man was supported entirely by the loop beneath the frame. In this case the strength of the arms as well as of the legs was applied, the first through the cranks and the latter through the little cross-piece at the tail. Both actions were conveyed to the wings through the same arrangement of cords. Since the operator's hands and feet were occupied, Leonardo proposed that the head should do the steering. This he would accomplish by inserting the head through a small loop to which was attached a long rudder tail, as indicated by the little sketch at the side. The wings in both devices were of the bat type. In whatever manner the machine was to be operated, Leonardo appreciated the fact that a nice balance, and one subject to easy adjustment, would be needed at all times. To this end he recommended that "a man with wings should be free from the waist upward in order to balance himself as he does in a boat, so that his center of gravity and that of the machine may be able to balance and to change when necessity requires, according to the change in the center of resistance."

Although Leonardo gave his best thoughts and utmost ingenuity to the perfection of a flying machine, he had no false confidence, for he attached to his plans this note of caution: "Try this machine over a lake and tie around your waist a long leather bottle for holding wine, so that if you should fall, you will not drown."

At the time that he was studying the possibility of forward motion, he also considered vertical motion downward by means of a parachute, upward by a helicopter. His experiment with the resisting or sustaining power of wings led him to see that a man could fall safely through the air if he had sufficient surface to furnish partial support. This area he computed to be about 24 square feet for a man of ordinary weight, for he recorded that "if a man has a piece of closely woven canvas measuring 12 braccia square, he will be able to lower himself from any great height without danger."

The parachute, however, was one of Leonardo's designs that did not remain hidden,

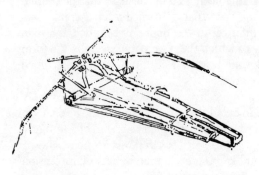

Figure 5A.

Figure 6.

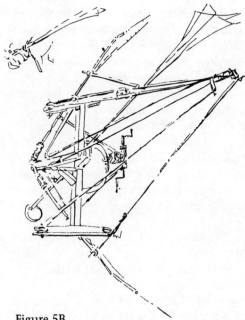

Figure 5B.

Figure 7.

for Verantius reproduced, or reinvented, it, when in his *Machinae novae* (?1595) he showed a man falling attached to a parachute (Fig. 6). Verantius gave his picture the title *Homo volans* (Man flying), indicating that the idea of aviation was in men's minds, although there was nothing in his design that suggested any mental conception of a flying machine. He was, therefore, a long way behind Leonardo.

In principle, the parachute is so simple, and so manifest to every boy who has flown a kite, that its application to the support of a man in slow descent did not call for great ingenuity on the part of anyone who could lay claim to powers of observation and deduction. In the converse proposition, that is, vertical upward flight, the matter is quite different, as then a problem of great difficulty is presented, one that has not yet been satisfactorily solved. The possibility of going straight up, something that a bird could not do, caught the fancy of Leonardo, and he was the first man to undertake the task of finding a way to accomplish it.

Like modern investigators, he adopted as his moving element the screw propeller, but in the form of a complete helix. This he did deliberately: he was aware that a bladed propeller would function, as he sketched one which, set in a chimney, revolved by the rise of the heated gases from the fire and turned a spit. In his description of his helicopter he based the action of the screw on an experiment with a wide and thin ruler. By "whirling it violently in the air," he said, "you will see that your arm will be guided by the line of the edge of said small board."

Leonardo sketched his helicopter (Fig. 7). The helix, which consisted of two complete turns, was made of linen whose pores had been sealed with a coating of starch; the edge was stiffened with an iron wire "as thick as a cord," and the frame was made of long and stout rods of wood. The helix had a diameter of eight braccia, so that it was no mean affair.

He ended his description with the conclusion that "if it is turned with speed the said screw will bore into the air and mount upward." Since reference is made to a model of cardboard, it is probable that in the experiments he succeeded in having a small screw go upward. Lacking a powerful motor capable of giving his large screw a very high rate of revolution, he could not have operated a heavy screw, particularly if he tried to lift his own weight in addition. Leonardo had the idea of wings and of a propeller and understood the action of air resistance, but his work stopped short of actual result, both in horizontal and in vertical flight, through the total absence of sufficient mechanical power. Not even his ingenuity could have forged, at that epoch, the single link that would have completed the chain.

As to actual attempts at flight, it is certain that he contemplated them, because in addition to his advice to use a machine over a lake and to carry a life preserver, he stated: "The first flight of the Big Bird will take place from the lofty Swan Hill [near Florence], and the universe will be filled with its praise and the nest whence it sprang will be filled with eternal glory."

Gerolamo Cardano (1501–1567), reviewing later trials, said that "Leonardo da Vinci had attempted to fly, but he was not successful."

For twenty-five years the top test pilot for Britain's Westland Aircraft, Harald Penrose here enacts Leonardo's dream, to soar and glide as a bird, to be its equal in every respect but the ability to flap one's wings. This passage from *I Flew With the Birds* (1949) relates a flight Penrose made over Dorset in 1936 in an open-cockpit sailplane of his own construction. His taste for the romance of flying has more recently been reflected in a massive historical study, *British Aviation: The Adventuring Years, 1920-29*.

HARALD PENROSE

I Flew With the Birds

The sunlight made the wings of the white gulls translucent as they soared across the summer blue. While assembling our long-winged sailplanes we watched the birds with interest, for they gave some idea of the extent of the up-currents on which presently we would fly.

At last my own machine was ready. It rested on the turf a hundred yards from the lichened stone wall which rambles along the top of the steep escarpment of Kimmeridge Hill. The scented wind sang in from the sea, two miles away, and leapt over the brow of the hill straight at the sailplane's silver nose, tugging gently at the held-down wings.

When the launching rope had been led from the quick release, around a distant pulley and so to the launching car, I wriggled carefully into the minute cockpit and fastened the safety belt. Soon the ground crew signaled all was ready. A wave of the hand, and the elastic catapult was steadily stretched. I shouted: "Release!" — and, with a smooth slide, the sailplane shot forward and lifted. Climbing steeply over the wall, into the powerful upsurge of wind deflected by the hill-slope, it freed from the catapult and gracefully made the slightest of dips, as if in salutation to the launching crew and earthbound things. The rushing wind embraced it buoyantly just as the sea holds a boat.

A light pressure on the controls, and the sailplane swung gracefully round until it was flying parallel with the hill ridge. In the strong up-current the machine climbed rapidly for a few moments. As the nose dropped into level flight the loveliness of an enchanting coast began to unfold. Etched against a vast expanse of shining sea the coast-line swept ruggedly from St. Aldhelm's Point, changing from shale to smooth curves of chalk as it stretched westward to Portland. Beyond the barrier of that headland the sea was fringed by the parallel curves of the Chesil Beach and the bright strip of the Fleet lake. But the eye swept further — far over the emptiness of West Bay to discover at last a glimpse of white which was Beer Head, the yellow of Sidmouth and Exmouth, and then the dim green of Tor Bay fading into the dark silhouette of Start Point, seventy miles away, below the dim purple shadows of Dartmoor.

Soaring steadily in the hill-wind the sailplane reached the far end of the ridge, canted steeply and returning, swept into a new vista of sunlit fields and moorland and the mirrored surface of the wide waterways of Poole. Like an animated shuttlecock in an infinity of space the sailplane began to sweep steadily to and fro, turning from the unending expanse of the blue Channel seas to the fair prospect of Dorset, and back again to the loveliness of sea.

The air was smooth as ice. With a strange and thrilling sibilance the wind whipped round the gleaming plywood of the cockpit, and curled over the little windscreen, flow-ing coldly on my face and stinging the eyes. It pressed firmly against the wings, holding the sailplane in cushioned arms, lifting it slowly higher and higher toward the gleam-ing cumulus that dotted the sky. It seemed that this gentle floating was magic; here at least was affinity with nature and harmony with space. The human shackles had gone — this was the unfettered flight of a bird. . . .

A bird — better than a bird surely? Higher, faster and quite as agile, I searched the hillside for the soaring gulls we had seen and at once found several, one group being only 200 feet below me. They gave the impression not of flying but of sliding along the hillside, their wings held arched and motionless. Slowly, very slowly, the sailplane closed on them — modern wings of polished ply and white fabric competing on equal terms with the gracefully feathered wings of ageless evolution.

Not all birds have the requisite wing shape to enable them to soar in the up-currents generated over the English countryside, but the long-winged gull is one of the greatest exponents. Along cliff edges in a blustering wind, skimming the steep thrown waves left by a winter gale, or high in the thermal currents of inland regions, gulls are found soaring, flying for miles with barely a wing-beat. Whether they evolved their slender wings to enable them to soar, or soar because they happen to have such wings may be the subject of fruitless debate, but the point of interest is that the aerodynamic char-acteristics of gulls are exactly matched to the light strength of the up-currents they most often meet. Had the bird a shorter wing, or been of much greater weight for the existing wing area, then soaring would have been impossible except in abnormal circumstances.

As I soared along that fragrant Dorset hillside, watching the gulls, I wondered again that natural laws enabled a mechanical creation to compete with such airy creatures. Wood, metal, and fabric, used in logical proportion and method, can give strength to withstand the wildest forces of the air and yet weigh, with their human load, no more than the wings can easily sustain by the suction of a moderate speed of air flowing over them. Thus my sailplane could be given wings of such a high ratio of overall span to wing area, yet bearing so moderate a total load, that the up-current required to nullify the machine's natural rate of descent was no greater than that required for the gull. In effect the sailplane dropped a little faster but on a flatter path than the bird. However, where the gull scored was in its ability to reef its wing area, and, by so doing, not only keep just ahead of me, but regulate its forward speed to either breeze or gale.

As I drew level with the birds I glanced at my airspeed indicator. The needle was steady at 35 miles an hour—a speed giving not my flattest glide, but a slow rate of descent. By going a little slower the sailplane's sinking speed became less than the uprising air, and so the machine began to climb. At 32 mph I seemed to be keeping station with the birds, but rising at under half a foot a second. I craned over the narrow cockpit side and watched the gulls intently.

There were three of them in a raggedly extended echelon, and as they flew they eyed sometimes the aerial scene and sometimes each other. Now and again a head would turn toward the wind-humming sailplane above them, and it was a coldly appraising yellow eye that watched. Though each gull was intensely sensitive to everything about it, with reactions set to a hair trigger, there was no fear of the giant wings above: the sailplane was a flying creature and that seemed sufficient for the bird.

All too soon I reached the ridge-end where I must turn, but the birds swung seaward, and, still holding height, began to traverse the airway above the half-mile of rough meadowland separating hill from shore. I waved farewell.

With nose well down, and a quartering wind to help, the sailplane swept back to the launching point and circled above the watching ground crew. Another gull flew by. I eased the control column a little forward, sensing the elevators pushing against the fluid air—and the sailplane steepened its glide. The wind hummed exultantly as the speed increased to 45 mph. Down and down went the sailplane. In a moment it was skimming the stunted bushes and bracken, slightly rising and falling in the rougher air just above the ground. Smoothly sliding, as though on ice, it raced to the downwind end of the hill ridge. As Swyre Head came into view I pulled up in a climbing turn almost on to the tail of a soaring herring gull.

Not more than 40 feet ahead, and a little higher than the glider, the gull sailed nonchalantly on. It glanced under a wing, then gave three lazy flaps. The impetus caused it to rise a few feet. I followed. The bird swept up a couple of yards. The sailplane closed a foot or two. With half a dozen wing beats the gull made good the difference, climbed a little higher—and settled to its soaring.

I turned the sailplane further from the hill face, so that our flight paths were parallel, but some fifty feet apart. Slowly the glider began to catch the bird. At 37 mph I was barely holding height. My speed dropped to 34. Much slower now, the distance between us lessened. Soon the bird was not more than ten feet ahead but on a course some way beyond my wing tip. It gave a few easy flaps and soared again. I rose on a gust. The bird climbed too, and then flapped for a moment to put itself well above me. From twenty feet higher it stared at me as we floated in formation past the ground crew, past the wind sock, past the last of the bracken, the beginning of the pasture, and then swept steadily around the long crescent of the hill.

"Cheat!" I called to the bird. "Stop flapping and I'll beat you at your own game." But the gull went sailing on, with an occasional wing beat that kept him master of the situation until presently we reached the end of my circuit. Like those others, the gull went soaring on toward the sea.

The next time I came to that point I also turned seawards. As though sailing a placid mill pool the glider went soaring away from the hill, holding its height just like the gull. Away and away from the up-currents of the hill it went, held up on a great mass of sun-warmed air rising from the fields between the cliffs and the hills. Cautiously exploring I began to find that in this area I could soar wherever I willed.

Presently I steered beyond the cliffs, nosing over the sea until the coastline lay half a mile behind and there seemed only blue around to the world's end, with the harsh rasp of the waves overpowering the gentle sound of the sailplane's flight.

In sudden apprehension I swept away from the endless water, speeding in a great curve back to the safety of the sun-warmed land. Kimmeridge Hill seemed far away — impossible to reach. But the gentle thermal still held the glider's wings, and the sea breeze was now a tail wind helping the race for home. Fields, trees, hedges, cows, flashed under the nose and were lost behind from sight. The ground began to rise. Suddenly the sailplane nosed into the strong up-current of the hill and rocketed above the summit, four hundred feet in half a minute.

Though I had intended to land, such a gift of height was too valuable to throw away at once, so four times more I soared along the hill, gazing at the loveliness of Wessex spreading wider and wider, until it was hidden in the purple shadows of the far horizons.

Reluctantly I turned the sailplane into the descending air behind the hilltop. On a smooth, heather-covered area my friends were ready to grasp the wings on landing. They stood motionless, their upturned faces white against the brown ridge.

Turning in smooth dropping curves, to left and right, the sailplane slid down its invisible air-slope toward the heather. One hundred feet up, fifty. Yes! going to land just by the crew! Twenty feet, ten, five, one — and with a fading sigh the sailplane lightly touched the ground and stopped.

For a moment the heather and long bracken, the distant shrubs, the tumbled stone wall marking the hill edge in front, were invested with an air of unreality. I looked up at the sky. White-winged, tranquil, effortless, a gull sailed by. Suddenly I was land-bound again, and I stared at the bird as though it were a creature of magic and only the earth was real.

Unlike Messrs. Abruzzo, Anderson, and Newman, the aeronauts who crossed the Atlantic in 1844 (from east to west, by the way) were quite insubstantial, being wholly the invention of Edgar Allan Poe. Passing references to such historic personages as George Cayley and W. S. Henson (inventor of the unsuccessful "aerial locomotive" in 1842) made more plausible such creations as Monck Mason, the Irish writer, musician, and aeronaut. Poe slipped this mélange of fact and fancy by the editors of the *New-York Sun*, who printed the "scoop" with screaming headlines on April 13, 1844.

EDGAR ALLAN POE

The Balloon-Hoax

THE BALLOON-HOAX!!!

[Astounding News! By Express *via* Norfolk! — The Atlantic Crossed in Three Days! Signal Triumph of Mr. Monck Mason's Flying Machine!!! — Arrival at Sullivan's Island, near Charleston, South Carolina, of Mr. Mason, Mr. Robert Holland, Mr. Henson, Mr. Harrison Ainsworth, and four others, in the Steering Balloon, "Victoria," after a Passage of Seventy-five Hours from Land to Land! Full Particulars of the Voyage!

The subjoined *jeu d'esprit* with the preceding heading in magnificent capitals, well interspersed with notes of admiration, was originally published, as matter of fact, in the *New-York Sun*, a daily newspaper, and therein fully subserved the purpose of creating indigestible aliment for the *quidnuncs* during the few hours intervening between a couple of the Charleston mails. The rush for the "sole paper which had the news," was something beyond even the prodigious; and, in fact, if (as some assert) the "Victoria" *did* not absolutely accomplish the voyage recorded, it will be difficult to assign a reason why she *should* not have accomplished it.]

The great problem is at length solved! The air, as well as the earth and the ocean, has been subdued by science, and will become a common and convenient highway for mankind. *The Atlantic has been actually crossed in a balloon!* and this too without difficulty — without any great apparent danger — with thorough control of the machine — and in the inconceivably brief period of seventy-five hours from shore to shore! By the energy of an agent at Charleston, South Carolina, we are enabled to be the first to furnish the public with a detailed account of this most extraordinary voyage, which was performed between Saturday, the 6th instant, at 11 A.M. and 2 P.M., on Tuesday, the 9th instant, by Sir Everard Bringhurst; Mr. Osborne, a nephew of Lord Bentinck's; Mr. Monck Mason and Mr. Robert Holland, the well-known aeronauts; Mr. Harrison Ainsworth, author of "Jack Sheppard," etc.; and Mr. Henson, the projector of the late unsuccessful flying machine — with two seamen from Woolwich — in all, eight persons. The particulars furnished below may be relied on as authentic and accurate in every respect, as, with a slight exception, they are copied *verbatim* from the joint diaries of Mr. Monck Mason and Mr. Harrison Ainsworth, to whose politeness our agent is also indebted for much verbal information respecting the balloon itself, its construction, and other matters of interest. The only alteration in the MS received, has been made for the purpose of throwing the hurried account of our agent, Mr. Forsyth, into a connected and intelligible form.

THE BALLOON

Two very decided failures, of late — those of Mr. Henson and Sir George Cayley — had much weakened the public interest in the subject of aerial navigation. Mr. Henson's scheme (which at first was considered very feasible even by men of science) was founded upon the principle of an inclined plane, started from an eminence by an extrinsic force, applied and continued by the revolution of impinging vanes, in form and number resembling the vanes of a windmill. But, in all the experiments made with models at the Adelaide Gallery, it was found that the operation of these fans not only did not propel the machine, but actually impeded its flight. The only propelling force it ever exhibited, was the mere *impetus* acquired from the descent of the inclined plane; and this *impetus* carried the machine farther when the vanes were at rest, than when they were in motion — a fact which sufficiently demonstrates their inutility; and in the absence of the propelling, which was also the *sustaining*, power, the whole fabric would necessarily descend. This consideration led Sir George Cayley to think only of adapting a propeller to some machine having of itself an independent power of support — in a word, to a balloon; the idea, however, being novel, or original, with Sir George, only so far as regards the mode of its application to practice. He exhibited a model of his invention at the Polytechnic Institution. The propelling principle, or power, was here, also, applied to interrupted surfaces, or vanes, put in revolution. These vanes were four in number, but were found entirely ineffectual in moving the balloon, or in aiding its ascending power. The whole project was thus a complete failure.

It was at this juncture that Mr. Monck Mason (whose voyage from Dover to Weilburg in the balloon, *Nassau*, occasioned so much excitement in 1837) conceived the idea of employing the principle of the Archimedean screw for the purpose of propulsion through the air—rightly attributing the failure of Mr. Henson's scheme, and of Sir George Cayley's, to the interruption of surface in the independent vanes. He made the first public experiment at Willis's Rooms, but afterward removed his model to the Adelaide Gallery ... where it accomplished a velocity of five miles per hour; although, strange to say, it excited very little interest in comparison with the previous complex machine of Mr. Henson—so resolute is the world to despise any thing which carries with it an air of simplicity. To accomplish the great desideratum of aerial navigation, it was very generally supposed that some exceedingly complicated application must be made of some unusually profound principle in dynamics.

So well satisfied, however, was Mr. Mason of the ultimate success of his invention, that he determined to construct immediately, if possible, a balloon of sufficient capacity to test the question by a voyage of some extent—the original design being to cross the British Channel, as before, in the *Nassau* balloon. To carry out his views, he solicited and obtained the patronage of Sir Everard Bringhurst and Mr. Osborne, two gentlemen well known for scientific acquirement, and especially for the interest they have exhibited in the progress of aerostation. The project, at the desire of Mr. Osborne, was kept a profound secret from the public. . . .

As the original design was to cross the British Channel, and alight as near Paris as possible, the voyagers had taken the precaution to prepare themselves with passports directed to all parts of the Continent, specifying the nature of the expedition, as in the case of the *Nassau* voyage, and entitling the adventurers to exemption from the usual formalities of office; unexpected events, however, rendered these passports superfluous.

"The inflation was commended very quietly at daybreak, on Saturday morning, the 6th instant, in the courtyard of Wheal-Vor House, Mr. Osborne's seat, about a mile from Penstruthal, in North Wales; and at seven minutes past eleven, everything being ready for departure, the balloon was set free, rising gently but steadily, in a direction nearly south; no use being made, for the first half hour, of either the screw or the rudder. We proceed now with the journal, as transcribed by Mr. Forsyth from the joint MSS of Mr. Monck Mason and Mr. Ainsworth. . . .

THE JOURNAL

Saturday, April the 6th.—Every preparation likely to embarrass us having been made overnight, we commenced the inflation this morning at daybreak; but owing to a thick fog, which encumbered the folds of the silk and rendered it unmanageable, we did not get through before nearly eleven o'clock. Cut loose, then, in high spirits, and rose gently but steadily with a light breeze at north, which bore us in the direction of the British Channel. Found the ascending force greater than we had expected; and as we rose higher and so got clear of the cliffs, and more in the sun's rays, our ascent became very rapid. I did not wish, however, to lose gas at so early a period of the adventure, and so concluded to ascend for the present. We soon ran out our guide-rope; but even when we had raised it clear of the earth, we still went up very rapidly. The balloon was unusually steady, and looked beautiful. In about ten minutes after starting, the barometer indicated an altitude of 15,000 feet. The weather was remarkably fine, and the view of the subjacent country—a most romantic one when seen from any point—was now especially sublime. The numerous deep gorges presented the appearance of lakes, on account of the dense vapors with which they were filled, and the pinnacles and crags to the southeast, piled in inextricable confusion, resembling nothing so much as the giant cities of Eastern fable. We were rapidly approaching the mountains in the south, but our elevation was more than sufficient to enable us to pass them in safety. In a few minutes we soared over them in fine style; and Mr. Ainsworth, with the seamen, was surprised at their apparent want of altitude when viewed from the car, the tendency of great elevation in a balloon being to reduce inequalities of the surface below, to nearly a dead level. At half-past eleven still proceeding nearly south, we obtained our first view of the Bristol Channel; and, in fifteen minutes afterward, the line of breakers on the coast appeared immediately beneath us, and we were fairly out at sea. We now resolved to let off enough gas to bring our guide-rope, with the buoys affixed, into the water. This was immediately done, and we commenced a gradual descent. In about twenty minutes our first buoy dipped, and at the touch of the second soon afterward, we remained stationary as to elevation. We were all now anxious to test the efficiency of the rudder and screw, and we put them both in requisition forthwith, for the purpose of altering our direction more to the eastward, and in a line for Paris. By means of the rudder we instantly effected the necessary change of direction, and our course was brought nearly at right angles to that of the wind; when we set in motion the spring of the screw, and were rejoiced to find it propel us readily as desired. Upon this we gave nine hearty cheers, and dropped in the sea a bottle, enclosing a slip of parchment with a brief account of the principle of the invention. Hardly, however, had we done with our rejoicings, when an unforeseen accident occurred which discouraged us in no little degree. The steel rod connecting the spring with the propeller was suddenly jerked out of place, at the car end (by a swaying of the car through some movement of one of the two seamen we had taken up), and in an instant hung dangling out of reach, from the pivot of the axis of the screw. While we were endeavoring to regain it, our attention being completely absorbed, we became involved in a strong current of wind from the east, which bore us, with rapidly increasing force, toward the Atlantic. We soon found ourselves driving out to sea at the rate of not less, certainly, than fifty or sixty miles an hour, so that we came up with Cape Clear, at some forty miles to our north, before we had secured the rod, and had time to think what we were about. It was now that Mr. Ainsworth made an extraordinary but, to my fancy, a by no means unreasonable or chimerical proposition, in which he was instantly seconded by Mr. Holland—viz.: that we should take advantage of the strong gale which bore us on, and in place of beating back to Paris, make an attempt to reach the coast of North America. . . . It is needless to say that a very short time sufficed us to lose sight of the coast. . . .

We kept on in this manner throughout the day with no material incident, and, as the shades of night closed around us, we made a rough estimate of the distance traversed. It could not have been less than 500 miles, and was probably much more. The propeller was kept in constant operation, and, no doubt, aided our progress materially. As the sun went down, the gale freshened into an absolute hurricane, and the ocean beneath was clearly visible on account of its phosphorescence. The wind was from the east all night, and gave us the brightest omen of success. We suffered no little from cold, and the dampness of the atmosphere was most unpleasant; but the ample space in the car enabled us to lie down, and by means of cloaks and a few blankets we did sufficiently well. . . .

Sunday, the 7th This morning the gale, by ten, had subsided to an eight- or nine-knot breeze (for a vessel at sea), and bears us, perhaps, thirty miles per hour, or more. It has veered, however, very considerably to the north; and now, at sundown, we are holding our course due west, principally by the screw and rudder, which answer their purposes to admiration. I regard the project as thoroughly successful, and the easy navigation of the air in any direction (not exactly in the teeth of a gale) as no longer problematical. We could not have made head against the strong wind of yesterday; but, by ascending, we might have got out of its influence, if requisite. Against a pretty stiff breeze, I feel convinced, we can make our way with the propeller. At noon, today, ascended to an elevation of nearly 25,000 feet, by discharging ballast. Did this to search for a more direct current, but found none so favorable as the one we are now in. We have an abundance of gas to take us across this small pond, even should the voyage last three weeks. I have not the slightest fear for the result. The difficulty has been strangely exaggerated and misapprehended. I can choose my current, and should I find *all* currents against me, I can make very tolerable headway with the propeller. We have had no incidents worth recording. The night promises fair. . . .

Monday, the 8th This morning we had again some little trouble with the rod of the propeller, which must be entirely remodeled, for fear of serious accident—I mean the steel rod, not the vanes. The latter could not be improved. The wind has been blowing steadily and strongly from the northeast all day; and so far fortune seems bent upon favoring us. Just before day, we were all somewhat alarmed at some odd noises and concussions in the balloon, accompanied with the apparent rapid subsidence of the whole machine. These phenomena were occasioned by the expansion of the gas, through increase of heat in the atmosphere, and the consequent disruption of the minute particles of ice with which the network had become encrusted during the night. Threw down several bottles to the vessels below. See one of them picked up by a large ship—seemingly one of the New York line packets. Endeavored to make out her name, but could not be sure of it. Mr. Osborne's telescope made it out something like *Atalanta*. It is now twelve at night, and we are still going nearly west, at a rapid pace. The sea is peculiarly phosphorescent. . . .

Tuesday, the 9th. . . . One P.M. *We are in full view of the low coast of South Carolina.* The great problem is accomplished. We have crossed the Atlantic—fairly and *easily* crossed it in a balloon! God be praised! Who shall say that any thing is impossible hereafter?

The journal here ceases. Some particulars of the descent were communicated, however, by Mr. Ainsworth to Mr. Forsyth. It was nearly dead calm when the voyagers first came in view of the coast. . . . [Mr. Osborne] having acquaintances at Fort Moultrie, it was immediately resolved to descend in its vicinity. The balloon was brought over the beach (the tide being out and the sand hard, smooth, and admirably adapted for a descent), and the grapnel let go, which took firm hold at once. The inhabitants of the island, and of the fort, thronged out, of course, to see the balloon; but it was with the greatest difficulty that any one could be made to credit the actual voyage—*the crossing of the Atlantic.* The grapnel caught at two P.M. precisely; and thus the whole voyage was completed in seventy-five hours; or rather less, counting from shore to shore. No serious accident occurred. No real danger was at any time apprehended. The balloon was exhausted and secured without trouble; and when the MS from which this narrative is compiled was despatched from Charleston, the party were still at Fort Moultrie. Their further intentions were not ascertained; but we can safely promise our readers some additional information either on Monday or in the course of the next day, at furthest.

This is unquestionably the most stupendous, the most interesting, and the most important undertaking ever accomplished or even attempted by man. What magnificent events may ensue, it would be useless now to think of determining.

This most difficult of aerobatic maneuvers was first executed by Jimmy Doolittle but soon was made to seem almost routine by the nineteen consecutive loops recorded by stunt flier Tex Rankin. This piece was published in the September 1930 issue of *Popular Aviation*, the magazine still with us as *Flying*.

TEX RANKIN and PHIL LOVE

How to Do an Outside Loop

"How are outside loops made?" is a question frequently in the minds of those interested in aviation. Here's what J. G. "Tex" Rankin, who recently made nineteen consecutive loops, for a new record, has to say:

"There are no rules regarding the establishment of a record for outside loops, so when I set out to 'try my luck' the local contest committee, consisting of Captain French, Lieutenant Smith, Lieutenant Clark, and Vance Breese (who were appointed by the President of the Portland Chapter of the National Aeronautic Association), made up a set of rules covering such a contest.

"The rules in part read that the plane would have to start a dive from a level flying position without first stalling the plane. The motor could be on or off. The pilot would continue in the dive, pushing the stick forward until the plane was on its back, then continue pushing the stick forward until the plane came out level on top in a direction not more than thirty degrees off from the direction in which the dive was started, without stalling or falling off to either side, but continuing a straight flight. There should not be more than two minutes' interval between each loop.

"Although I made thirty-four loops in a total elapsed time of fifty minutes without stalling or falling off on any of them, I was granted only nineteen by the committee. This was due to the fact that I came out slightly more than thirty degrees off on the top of fifteen of these loops, which was caused by the difficulty I experienced in keeping my feet on the rudder—there being no straps provided for this purpose.

"I found that my best loops were made by starting from level flight with the stabilizer set clear down to make the ship nose-heavy, with the engine wide open, and making a dive of 3,000 feet around an arc of 1,500 feet, and coming up the other side in the same manner. My net loss on altitude was about 500 feet. I also found that if the ship was pulled into a stall before starting the dive that it usually resulted in a sloppy loop. This also occurred if I did not allow the plane to come up high enough before leveling off. In other words, if I dived 3,000 feet and tried to level off with a new loss of 1,500 feet in altitude, it would cause the ship to roll on top of the loop. I experimented on outside loops for several weeks before attempting to establish a record, gaining much useful information from these experimental flights.

"I carried a parachute, but no other safety belt than the regular standard one that was already provided on the ship. I experienced nothing unusual while doing these loops, and I feel that anyone with sufficient flying experience—and particularly with experience in inverted flight—would have no difficulty in making as many outside loops as I did, or even more, providing of course that he be in good average physical condition and use a strong, well-made ship."

The outside loop is probably the most exciting and hazardous stunt known to flying. It is so called because the pilot is on the outer rim of the loop during the maneuver instead of on the inner side as in the ordinary loop.

Lieutenant Phil Love, flying partner of Colonel Lindbergh's on the airmail run, has the honor of being the first to do an outside loop with a passenger. The plane used was a tapered wing Waco, powered with a J-6 seven-cylinder motor, but the ship was not quite standard in every respect. An auxiliary tank was built into the landing gear, in order to furnish gravity flow to the carburetor during inverted flight. In normal flight this tank is filled by gravity from the main tank but when flight is inverted a check valve stops the flow back to the main tank, allowing the gasoline to run only to the carburetor. All of the oil breathers are taped up and the main tank is vented through a long pipe to the landing gear so that when inverted the top of the vent is above the main gas-tank level.

Shortly after landing from the loops Phil gave a report on his experiences. He said that there was very little experience while actually undergoing the loop itself.

"I have experimented with several different ways of executing them," he continued, "I have tried stalling into the initial dive both with power on and with power off, and with different stabilizer settings, but I find that the most satisfactory way is to set the stabilizer very slightly nose-heavy, and to fly the ship into the dive with power on from cruising speed.

"It takes quite a bit of determination to shove the stick forward until the nose is straight down, and then keep shoving it forward until the ship is horizontal on its back. The execution of the remaining portion of the loop is the most difficult, for after the nose has risen above the horizon there is no reference point to level the ship by, and it is at this point that the pressure tending to throw the pilot out of the cockpit is greatest. In addition to this centrifugal pressure which is aided by gravity it is necessary to apply no little amount of forward pressure on the stick to force the ship up and over the top.

"I find that going up the back side of the loop the ship has a tendency to rotate and care must be exercised to prevent this. The horizon is visible out of both sides of the cockpit, but I have never been able to tell much about the attitude of the ship by looking at it in this position. I find it more satisfactory to make sure that the ship is absolutely level at the bottom of the loop while the horizon is visible and relative to the wings, and then to shove the stick exactly straight forward to prevent turning with the ailerons, and carrying just enough rudder to counteract 'torque.' If this is done correctly the chances are that the loop will be completed with the wings level, and the ship flying in the same line as when the loop was started.

"The Waco taper wing does its best loops starting with power on at about 100 miles an hour and about 140 miles an hour at the bottom. The diameter of the loop is about 1,000 feet, and very little altitude, if any, is lost on completion of the maneuver.

"After completing several loops in close succession I found that the rush of blood to the head causes an extremely warm feeling about the temples, and if one isn't in good condition he is apt to be a little bit 'woozy,' but this soon passes. Ordinarily upon landing the pilot will find that his eyes are slightly bloodshot, but no aftereffects are noticed unless too much speed was gained in the dive, and an attempt to change direction too fast was made. It is this changing direction snappily that throws the tremendously excessive strain on both pilot and ship."

During Lieutenant Love's flight an ordinary safety belt was used, the only added precaution being a heavy rubber band to hold the clasp from opening accidentally.

The Red Baron registered eighty official victories, tops among aces of all nations, before he and his red Fokker triplane were shot down in April 1918. But Rittmeister Manfred von Richtofen had made his mark on history one year before, in the "Bloody April" of 1917, when the superiority of German airplanes—particularly the Albatros D.III, which the Baron flew at that time— brought devastating losses upon the airmen of Britain. In that month alone, Germany bagged 150 British planes, and Richtofen upped his personal count from thirty-one to fifty-two. This passage is from his autobiography, translated by T. Ellis Barker.

MANFRED VON RICHTOFEN

April, Bloody April

MY FIRST DOUBLE EVENT

The second of April, 1917, was a very warm day for my squadron. From my quarters I could clearly hear the drum-fire of the guns which was again particularly violent.

I was still in bed when my orderly rushed into the room and exclaimed: "Sir, the English are here!" Sleepy as I was, I looked out of the window and, really, there were my dear friends circling over the flying ground. I jumped out of my bed and into my clothes in a jiffy. My Red Bird had been pulled out and was ready for starting. My mechanics knew that I should probably not allow such a favorable moment to go by unutilized. Everything was ready. I snatched up my furs and then went off.

I was the last to start. My comrades were much nearer to the enemy. I feared that my prey would escape me, that I should have to look on from a distance while the others were fighting. Suddenly one of the impertinent fellows tried to drop down upon me. I allowed him to come near and then we started a merry quadrille. Sometimes my opponent flew on his back and sometimes he did other tricks. He had a double-seated chaser. I was his master and very soon I recognized that he could not escape me.

During an interval in the fighting I convinced myself that we were alone. It followed that the victory would accrue to him who was calmest, who shot best, and who had the clearest brain in a moment of danger. After a short time I got him beneath me without seriously hurting him with my gun. We were at least two kilometers from the front. I thought he intended to land but there I had made a mistake. Suddenly, when he was only a few yards above the ground, he once more went off on a straight course. He tried to escape me. That was too bad. I attacked him again and I went so low that I feared I should touch the roofs of the houses of the village beneath me. The Englishman defended himself up to the last moment. At the very end I felt that my engine had been hit. Still I did not let go. He had to fall. He rushed at full speed right into a block of houses.

There was little left to be done. This was once more a case of splendid daring. He defended himself to the last. However, in my opinion he showed more foolhardiness than courage. This was one of the cases where one must differentiate between energy and idiocy. He had to come down in any case but he paid for his stupidity with his life.

I was delighted with the performance of my red machine during its morning work and returned to our quarters. My comrades were still in the air and they were very surprised, when, as we met at breakfast, I told them that I had scored my thirty-second machine.

A very young lieutenant had "bagged" his first airplane. We were all very merry and prepared everything for further battles.

I then went and groomed myself. I had not had time to do it previously. I was visited by a dear friend, Lieutenant Voss of Boelcke's squadron. We chatted. Voss had downed on the previous day his twenty-third machine. He was next to me on the list and is at present my most redoubtable competitor.

When he started to fly home I offered to accompany him part of the way. We went on a roundabout way over the fronts. The weather had turned so bad that we could not hope to find any more game.

Beneath us there were dense clouds. Voss did not know the country and he began to feel uncomfortable. When we passed above Arras I met my brother who also is in my squadron and who had lost his way. He joined us. Of course he recognized me at once by the color of my machine.

Suddenly we saw a squadron approaching from the other side. Immediately the thought occurred to me: "Now comes number thirty-three." Although there were nine Englishmen and although they were on their own territory they preferred to avoid battle. I thought that perhaps it would be better for me to repaint my machine. Nevertheless we caught them up. The important thing in airplanes is that they are speedy.

I was nearest to the enemy and attacked the man to the rear. To my greatest delight I noticed that he accepted battle and my pleasure was increased when I discovered that his comrades deserted him. So I had once more a single fight.

It was a fight similar to the one which I had had in the morning. My opponent did not make matters easy for me. He knew the fighting business and it was particularly awkward for me that he was a good shot. To my great regret that was quite clear to me.

BARON MANFRED VON RICHTHOFEN,

GERMANY'S FABLED

RED BARON

THE LEADING ACE OF WORLD WAR I, CRASHED ON HIS FIRST SOLO FLIGHT.

UNDAUNTED, HE WENT ON TO SHOOT DOWN 80 ALLIED PLANES.

ON APRIL 21, 1918, CANADIAN ROY BROWN FLYING IN THE R.A.F. ENDED THE BARON'S CAREER, SHOOTING HIM DOWN NEAR AMIENS, FRANCE.

A favorable wind came to my aid. It drove both of us into the German lines.* My opponent discovered that the matter was not so simple as he had imagined. So he plunged and disappeared in a cloud. He had nearly saved himself.

I plunged after him and dropped out of the cloud and, as luck would have it, found myself close behind him. I fired and he fired without any tangible result. At last I hit him. I noticed a ribbon of white benzine vapor. He had to land for his engine had come to a stop.

He was a stubborn fellow. He was bound to recognize that he had lost the game. If he continued shooting I could kill him, for meanwhile we had dropped to an altitude of about nine hundred feet. However, the Englishman defended himself exactly as did his countryman in the morning. He fought until he landed. When he had come to the ground I flew over him at an altitude of about thirty feet in order to ascertain whether I had killed him or not. What did the rascal do? He took his machine gun and shot holes into my machine.

Afterwards Voss told me if that had happened to him he would have shot the airman on the ground. As a matter of fact I ought to have done so for he had not surrendered. He was one of the few fortunate fellows who escaped with their lives.

I felt very merry, flew home and celebrated my thirty-third airplane.

MY RECORD-DAY

The weather was glorious. We were ready for starting. I had as a visitor a gentleman who had never seen a fight in the air or anything resembling it and he had just assured me that it would tremendously interest him to witness an aerial battle.

We climbed into our machines and laughed heartily at our visitor's eagerness. Friend Schäfer† thought that we might give him some fun. We placed him before a telescope and off we went.

The day began well. We had scarcely flown to an altitude of six thousand feet when an English squadron of five machines was seen coming our way. We attacked them by a rush as if we were cavalry and the hostile squadron lay destroyed on the ground.

---

*It is well to note how often von Richthofen refers to the wind being in his favor. A west wind means that while the machines are fighting they are driven steadily over the German lines. Then, if the British machine happens to be inferior in speed or maneuverability to the German, and is forced down low, the pilot has the choice only of fighting to a finish and being killed, or of landing and being made prisoner. The prevalence of west winds has, for this reason, cost the RFC a very great number of casualties in killed and missing, who, if the fight had occurred over territory held by the British, would merely have landed till the attacking machine had taken itself off. For similar reasons, the fact that the RFC has always been on the offensive, and so has always been flying over the German lines has caused many casualties. Under all the circumstances it is surprising that the RFC casualties have not been a great deal heavier.

†Schäfer was also shot by Lieutenant Rhys-Davids, RFC, later in 1917.

None of our men was even wounded. Of our enemies three had plunged to the ground and two had come down in flames.

The good fellow down below was not a little surprised. He had imagined that the affair would look quite different, that it would be far more dramatic. He thought the whole encounter had looked quite harmless until suddenly some machines came falling down looking like rockets. I have gradually become accustomed to seeing machines falling down, but I must say it impressed me very deeply when I saw the first Englishman fall and I have often seen the event again in my dreams.

As the day had begun so propitiously we sat down and had a decent breakfast. All of us were as hungry as wolves. In the meantime our machines were again made ready for starting. Fresh cartridges were got and then we went off again.

In the evening we could send off the proud report: "Six German machines have destroyed thirteen hostile airplanes."*

Boelcke's squadron had only once been able to make a similar report. At that time we had shot down eight machines. Today one of us had brought low four of his opponents. The hero was a Lieutenant Wolff, a delicate-looking little fellow in whom nobody could have suspected a redoubtable hero. My brother had destroyed two, Schäfer two, Festner two, and I three.

We went to bed in the evening tremendously proud but also terribly tired. On the following day we read with noisy approval about our deeds of the previous day in the official communique. On the next day we downed eight hostile machines.

A very amusing thing occurred. One of the Englishmen whom we had shot down and whom we had made a prisoner was talking with us. Of course he inquired after the Red Airplane. It is not unknown even among the troops in the trenches and is called by them *le diable rouge*. In the squadron to which he belonged there was a rumor that the Red Machine was occupied by a girl, by a kind of Jeanne d'Arc. He was intensely surprised when I assured him that the supposed girl was standing in front of him. He did not intend to make a joke. He was actually convinced that only a girl could sit in the extravagantly painted machine.

*It is possible that the figures are correct. Early in 1917, before the advent of the British fighters and de Havilands in quantities, the RFC was having a very bad time. On April 7, for example, it was reported in the GHQ Communique that twenty-eight English machines were missing.

When Victor Chapman was shot down behind enemy lines in 1916 while fighting five Fokkers, he became the first man to die in an American fighting unit of World War I. An impulsive, daring pilot who took perhaps too little heed of his own safety, Chapman fell to his death with his head in bandages from a near escape of only the week before. His comrade in the Lafayette Escadrille, Kiffin Rockwell, struck back by quickly downing four German planes, but he, too, was to meet his death before the year was out.

KIFFIN ROCKWELL

"He Died the Most Glorious Death"

ESCADRILLE N. 124, SECTEUR 24
AUGUST 10, 1916

My dear Mrs. Chapman

I received your letter this morning. I feel mortified that you have had to write me without my having written you before, when Victor was the best friend I ever had. I wanted to write you and his father at once, and tried to a number of times. But I found it impossible to write full justice to Victor or to really express my sympathy with you. Everything I would try to say seemed so weak. So I finally said, "I will just go ahead and work hard, do my best, then if I have accomplished a lot or been killed in accomplishing it, they will know that I had not forgotten Victor, and that some of his strength of character still lived." There is nothing that I can say to you or anyone that will do full credit to him. And everyone here that knew him feels the same way. To start with, Victor had such a strong character. I think we all have our ideals when we begin but unfortunately there are so very few of us that retain them; and sometimes we lose them at a very early age and after that, life seems to be spoiled. But Victor was one of the very few who had the strongest of ideals, and then had the character to withstand anything that tried to come into his life and kill them. He was just a large, healthy man, full of life and goodness toward life, and could only see the fine, true points in life and in other people. And he was not of the kind that absorbs from other people, but of the kind that gives out. We all had felt his influence, and seeing in him a man, made us feel a little more like trying to be men ourselves.

When I am in Paris, I stay with Mrs. Weeks, whose son was my friend, and killed in the Legion. Well, Victor would come around once in a while to dinner with us. Mrs. Weeks used always to say to me: "Bring Victor around, he does me so much good. I like his laugh and the sound of his voice. When he comes in the room it always seems so much brighter." Well, that is the way it was here in the Escadrille.

For work in the Escadrille, Victor worked hard, always wanting to fly. And courage! he was too courageous, we all would beg him at times to slow up a little. We speak of him every day here, and we have said sincerely amongst ourselves many a time that Victor had more courage than all the rest of the Escadrille combined. He would attack the Germans always, no matter what the conditions or what the odds. The day he was wounded, four or five of the Escadrille had been out and come home at the regular hour. Well, Victor had attacked one machine and seriously crippled it, but the machine has succeeded in regaining the German lines. After that Victor would not come home with the rest but stayed looking for another machine. He found five machines inside our lines. None of us like to see a German machine within our lines, without attacking. So, although Victor was alone, he watched the five machines and finally one of them came lower and under him. He immediately dived on this one. Result was that the others dived on him. One of them was a Fokker, painted like the machine of the famous Captain Boelcke and may have been him. This Fokker got the position on Victor, and it was a miracle that he was not killed then. He placed bullet after bullet around Victor's head, badly damaging the machine, cutting parts of the command in two, and one bullet cutting his scalp, as you know. Well, Victor got away, and with one hand held the commands together where they had been cut and landed at Froids where we had friends in a French Escadrille. There he had dinner and his wound was dressed, and they repaired his machine a little. That afternoon he came flying back home with his head all bound up. Yet he thought nothing of it, only smiled and thought it an interesting event. He immediately wanted to continue his work as if nothing had happened. We tried to get him to go to a hospital, or to go to Paris for a short while and rest; but he said no. Then we said, "Well you have got to take a rest, even if you stay here." The Captain told him that he would demand a new and better machine for him, and that he could rest while waiting for it to be ready, and then could see whether or not he should go back to flying. This was the 17th of June. The following morning Balsley was wounded. The same day or the day after, Uncle Willie came to see Victor and was with us a couple of days. Those first days Victor slept late, a privilege he had not taken before since being in the Escadrille, always having got up at daylight. In the daytime he would be with Uncle Willie, or at the field, seeing about his machine, or he would take his old machine and fly over to see Balsley. At first Balsley could not eat or drink anything. But after a few days he was allowed a little champagne and oranges. Well, as soon as Victor found that out, he arranged for champagne to be sent to Balsley, and would take oranges over to him. At least once a day, and sometimes twice, he would go over to see Balsley to cheer him up. And in the meantime he wouldn't ever let anyone speak of his wound, as a wound, and was impatient for his new machine. On the 21st he got his machine and had it regulated. On the 22nd he regulated the mitrailleuse, and the weather being too bad to fly over the lines, he flew it around here a little to get used to it. His head was still bandaged, but he said it was nothing. Late in the afternoon some German machines were signaled and he went up with the rest of us to look for them, but it was a false alarm. The following morning the weather was good, and he insisted on going out at the regular hour with the rest. There were no machines over the lines, so the *sortie* was uneventful. He came in, and at lunch fixed up a basket of oranges which he said he would take to Balsley. We went up to the field, and Captain Thenault, Prince and Lufberry got ready to go out and over the lines. Victor put the oranges in his machine and said that he would follow the others over the lines for a little trip and then go and land at the hospital. The Captain, Prince, and Lufberry started first. On arriving at the lines they saw at first two German machines which they dived on. When they arrived in the midst of them, they found that two or three other German machines had arrived also. As the odds were against the three, they did not fight long, but immediately started back into our lines and without seeing Victor. When they came back we thought that Victor was at the hospital. But later in the afternoon a *pilote* of a Maurice Farman and his passenger sent in a report. The report was that they saw three Nieuports attack five German machines, that at this moment they saw a fourth Nieuport arriving with all speed who dived in the midst of the Germans, that two of the Germans dived toward their field and that the Nieuport fell through the air no longer controlled by the *pilote*. In a fight it is practically impossible to tell what the other machines do, as everything happens so fast and all one can see is the beginning of a fight and then, in a few seconds, the end. That fourth Nieuport was Victor and, owing to the fact that the motor was going at full speed when the machine fell, I think that he was killed instantly.

He died the most glorious death, and at the most glorious time of life to die, especially for him with his ideals. I have never once regretted it for him, as I know he was willing and satisfied to give his life that way if it was necessary, and that he had no fear of death, and there is nothing to fear in death. It is for you, his father, relatives, myself, and for all who have known him, and all who would have known him, and for the world as a whole I regret his loss. Yet he is not dead, he lives forever in every place he has been, and in everyone who knew him, and in the future generations little points of his character will be passed along. He is alive every day in this Escadrille and has a tremenduous influence on all our actions. Even the *mécaniciens* do their work better and more conscientiously. And a number of times I have seen Victor's *mécanicien* standing (when there was no work to be done) and gazing off in the direction of where he last saw Victor leaving for the lines.

For promotions and decorations things move slowly in the army, and after it has passed through all the bureaus, it takes some time to get back to you. Victor was proposed for sergeant and for the Croix de Guerre May 24th. This passed through all the bureaus and was signed by the general, but the papers did not arrive here until June 25th. However, Victor knew on the 23rd, that they had passed, and that it was only a question of a day or so. He had also been promised, after being wounded, the *Médaille Militaire* which he would have received sometime in July. I wish that they could have sent that to you, for he had gained it, and they would have given it to him. But it is against the rules to give the *Médaille Militaire* unless everything has been signed before the *titulaire* is killed.

I must close now. You must not feel sorry, but must feel proud and happy.

Every flier has the soul of a poet: Although he may not express his feelings in poetic terms, he confronts nature and experiences beauty in a way beyond the ken of earthbound man. Antoine de Saint-Exupéry gave voice to those feelings in his unique and enduring *Wind, Sand and Stars*, an impressionistic evocation, in the words of André Gide, of "the sensations, emotions and reflections of the airman." Not all of that 1939 classic, however, is given over to speculative flights into the purple — the sturdy, direct passage below is the very opening of the book.

ANTOINE DE SAINT-EXUPERY

Below the Sea of Clouds Lies Eternity

In 1926 I was enrolled as student airline pilot by the Latécoère Company, the predecessors of Aéropostale (now Air France) in the operation of the line between Toulouse, in southwestern France, and Dakar, in French West Africa. I was learning the craft, undergoing an apprenticeship served by all young pilots before they were allowed to carry the mails. We took ships up on trial spins, made meek little hops between Toulouse and Perpignan, and had dreary lessons in meteorology in a freezing hangar. We lived in fear of the mountains of Spain, over which we had yet to fly, and in awe of our elders.

These veterans were to be seen in the field restaurant — gruff, not particularly approachable, and inclined somewhat to condescension when giving us the benefit of their experience. When one of them landed, rain-soaked and behind schedule, from Alicante or Casablanca, and one of us asked humble questions about his flight, the very curtness of his replies on these tempestuous days was matter enough out of which to build a fabulous world filled with snares and pitfalls, with cliffs suddenly looming out of fog and whirling air currents of a strength to uproot cedars. Black dragons guarded the mouths of the valleys and clusters of lightning crowned the crests — for our elders were always at some pains to feed our reverence. But from time to time one or another of them, eternally to be revered, would fail to come back.

I remember, once, a homecoming of Bury, he who was later to die in a spur of the Pyrenees. He came into the restaurant, sat down at the common table, and went stolidly at his food, shoulders still bowed by the fatigue of his recent trial. It was at the end of one of those foul days when from end to end of the line the skies are filled with dirty weather, when the mountains seem to a pilot to be wallowing in slime like exploded cannon on the decks of an antique man-o'-war.

I stared at Bury, swallowed my saliva, and ventured after a bit to ask if he had had a hard flight. Bury, bent over his plate in frowning absorption, could not hear me. In those days we flew open ships and thrust our heads out round the windshield, in bad weather, to take our bearings: the wind that whistled in our ears was a long time clearing out of our heads. Finally Bury looked up, seemed to understand me, to think back to what I was referring to, and suddenly he gave a bright laugh. This brief burst of laughter, from a man who laughed little, startled me. For a moment his weary being was bright with it. But he spoke no word, lowered his head, and went on chewing in silence. And in that dismal restaurant, surrounded by the simple government clerks who sat there repairing the wear and tear of their humble daily tasks, my broad-shouldered messmate seemed to me strangely noble; beneath his rough hide I could discern the angel who had vanquished the dragon.

The night came when it was my turn to be called to the field-manager's room.

He said: "You leave tomorrow."

I stood motionless, waiting for him to dismiss me. After a moment of silence he added:

"I take it you know the regulations?"

In those days the motor was not what it is today. It would drop out, for example, without warning and with a great rattle like the crash of crockery. And one would simply throw in one's hand: there was no hope of refuge on the rocky crust of Spain. "Here," we used to say, "when your motor goes, your ship goes, too."

An airplane, of course, can be replaced. Still, the important thing was to avoid a collision with the range; and blind flying through a sea of clouds in the mountain zones was subject to the severest penalties. A pilot in trouble who buried himself in the white cotton wool of the clouds might all unseeing run straight into a peak. This was why, that night, the deliberate voice repeated insistently its warning:

"Navigating by the compass in a sea of clouds over Spain is all very well, it is very dashing, but — "

And I was struck by the graphic image:

"But you want to remember that below the sea of clouds lies eternity."

And suddenly that tranquil cloud-world, that world so harmless and simple that one sees below on rising out of the clouds, took on in my eyes a new quality. That peaceful world became a pitfall. I imagined the immense white pitfall spread beneath me. Below it reigned not what one might think — not the agitation of men, not the living tumult and bustle of cities, but a silence even more absolute than in the clouds, a peace even more final. This viscous whiteness became in my mind the frontier between the real and the unreal, between the known and the unknowable. Already I was beginning to realize that a spectacle has no meaning except it be seen through the glass of a culture, a civilization, a craft. Mountaineers too know the sea of clouds, yet it does not seem to them the fabulous curtain it is to me.

When I left that room I was filled with a childish pride. Now it was my turn to take on at dawn the responsibility of a cargo of passengers and the African mails. But at the same time I felt very meek. I felt myself ill-prepared for this responsibility. Spain was poor in emergency fields; we had no radio; and I was troubled lest when I got into difficulty I should not know where to hunt a landing place. Staring at the aridity of my maps, I could see no help in them; and so, with a heart full of shyness and pride, I fled to spend this night of vigil with my friend Guillaumet. Guillaumet had been over the route before me. He knew all the dodges by which one got hold of the keys to Spain. I should have to be initiated by Guillaumet.

When I walked in he looked up and smiled.

"I know all about it," he said. "How do you feel?"

He went to a cupboard and came back with glasses and a bottle of port, still smiling.

"We'll drink to it. Don't worry. It's easier than you think."

Guillaumet exuded confidence the way a lamp gives off light. He was himself later on to break the record for postal crossings in the Andes and the South Atlantic. On this night, sitting in his shirt-sleeves, his arms folded in the lamplight, smiling the most heartening of smiles, he said to me simply:

"You'll be bothered from time to time by storms, fog, snow. When you are, think of those who went through it before you, and say to yourself, 'What they could do, I can do.'"

I spread out my maps and asked him hesitantly if he would mind going over the hop with me. And there, bent over in the lamplight, shoulder to shoulder with the veteran, I felt a sort of schoolboy peace.

But what a strange lesson in geography I was given! Guillaumet did not teach Spain to me, he made the country my friend. He did not talk about provinces, or peoples, or livestock. Instead of telling me about Guadix, he spoke of three orange trees on the edge of the town: "Beware of those trees. Better mark them on the map." And those three orange trees seemed to me thenceforth higher than the Sierra Nevada.

He did not talk about Lorca, but about a humble farm near Lorca, a living farm with its farmer and the farmer's wife. And this tiny, this remote couple, living a thousand miles from where we sat, took on a universal importance. Settled on the slope of a mountain, they watched like lighthouse-keepers beneath the stars, ever on the lookout to succor men.

The details that we drew up from oblivion, from their inconceivable remoteness, no geographer had been concerned to explore. Because it washed the banks of great cities, the Ebro River was of interest to mapmakers. But what had they to do with that brook running secretly through the waterweeds to the west of Motril, that brook nourishing a mere score or two of flowers?

"Careful of that brook: it breaks up the whole field. Mark it on your map." Ah, I was to remember that serpent in the grass near Motril! It looked like nothing at all, and its faint murmur sang to no more than a few frogs; but it slept with one eye open. Stretching its length along the grasses in the paradise of that emergency landing field, it lay in wait for me a thousand miles from where I sat. Given the chance, it would transform me into a flaming candelabrum. And those thirty valorous sheep ready to charge me on the slope of a hill! Now that I knew about them I could brace myself to meet them.

"You think the meadow empty, and suddenly bang! there are thirty sheep in your wheels." An astounded smile was all I could summon in the face of so cruel a threat.

Little by little, under the lamp, the Spain of my map became a sort of fairyland. The crosses I marked to indicate safety zones and traps were so many buoys and beacons. I charted the farmer, the thirty sheep, the brook. And, exactly where she stood, I set a buoy to mark the shepherdess forgotten by the geographers.

When I left Guillaumet on that freezing winter night, I felt the need of a brisk walk. I turned up my coat collar, and as I strode among the indifferent passersby I was escorting a fervor as tender as if I had just fallen in love. To be brushing past these strangers with that marvelous secret in my heart filled me with pride. I seemed to myself a sentinel standing guard over a sleeping camp. These passersby knew nothing about me, yet it was to me that, in their mail pouches, they were about to confide the weightiest cares of their hearts and their trade. Into my hands were they about to entrust their hopes. And I, muffled up in my cloak, walked among them like a shepherd, though they were unaware of my solicitude.

Nor were they receiving any of those messages now being despatched to me by the night. For this snowstorm that was gathering, and that was to burden my first flight, concerned my frail flesh, not theirs. What could they know of those stars that one by one were going out? I alone was in the confidence of the stars. To me alone news was being sent of the enemy's position before the hour of battle. My footfall rang in a universe that was not theirs.

These messages of such grave concern were reaching me as I walked between rows of lighted shop windows, and those windows on that night seemed a display of all that was good on earth, of a paradise of sweet things. In the sight of all this happiness, I tasted the proud intoxication of renunciation. I was a warrior in danger. What meaning could they have for me, these flashing crystals meant for men's festivities, these lamps whose glow was to shelter men's meditations, these cozy furs out of which were to emerge pathetically beautiful solicitous faces? I was still wrapped in the aura of friendship, dazed a little like a child on Christmas Eve, expectant of surprise and palpitatingly prepared for happiness; and yet already I was soaked in spray; a mail pilot, I was already nibbling the bitter pulp of night flight.

It was three in the morning when they woke me. I thrust the shutters open with a dry snap, saw that rain was falling on the town, and got soberly into my harness. A half-hour later I was out on the pavement shining with rain, sitting on my little valise and waiting for the bus that was to pick me up. So many other flyers before me, on their day of ordination, had undergone this humble wait with beating heart.

Finally I saw the old-fashioned vehicle come around the corner and heard its tinny rattle. Like those who had gone before me, I squeezed in between the sleepy customs guard and a few glum government clerks. The bus smelled musty, smelled of the dust of government offices into which the life of a man sinks as into a quicksand. It stopped every five hundred yards to take on another scrivener, another guard, another inspector.

Those in the bus who had already gone back to sleep responded with a vague grunt to the greeting of the newcomer, while he crowded in as well as he was able and instantly fell asleep himself. We jolted mournfully over the uneven pavements of Toulouse, I in the midst of these men who in the rain and the breaking day were about to take up again their dreary diurnal tasks, their red tape, their monotonous lives.

Morning after morning, greeted by the growl of the customs guard shaken out of sleep by his arrival, by the gruff irritability of clerk or inspector, one mail pilot or another got into this bus and was for the moment indistinguishable from these bureaucrats. But as the street lamps moved by, as the field drew nearer and nearer, the old omnibus rattling along lost little by little its reality and became a grey chrysalis from which one emerged transfigured.

Morning after morning a flyer sat here and felt of a sudden, somewhere inside the vulnerable man subjected to his neighbor's surliness, the stirring of the pilot of the Spanish and African mails, the birth of him who, three hours later, was to confront in the lightnings the dragon of the mountains; and who, four hours afterwards, having vanquished it, would be free to decide between a detour over the sea and a direct assault upon the Alcoy range, would be free to deal with storm, with mountain, with ocean.

And thus every morning each pilot before me, in his time, had been lost in the anonymity of daybreak beneath the dismal winter sky of Toulouse, and each one, transfigured by this old omnibus, had felt the birth within him of the sovereign who, five hours later, leaving behind him the rains and snows of the North, repudiating winter, had throttled down his motor and begun to drift earthward in the summer air beneath the shining sun of Alicante.

By turns preposterous and prescient, this article appeared in the September 8, 1860, issue of *Scientific American*. Gliders, helicopters, airplanes, rockets — everything suddenly seemed possible to the publication's editors, who not long before had believed that the pursuit of human flight "contained no promise of success." The Mr. Hyatt mentioned in the second paragraph was Thaddeus Hyatt, who in that year offered a $1,000 prize for the best flying machine. Even in preinflated dollars, one would have thought such an invention might have been worth a bit more.

SCIENTIFIC AMERICAN

Flying Machines in the Future

Of all inventions of which it is possible to conceive in the future, there is none which so captivates the imagination as that of a flying machine. The power of rising up into the air and rushing in any direction desired at the rate of a mile or more in a minute is a power for which mankind would be willing to pay very liberally. What a luxurious mode of locomotion! To sweep along smoothly, gracefully, and swiftly over the tree-tops, changing the course at pleasure, and alighting at will. How perfectly it would eclipse all other means of travel by land and sea! This magnificent problem, so alluring to the imagination and of the highest practical convenience and value, has been left heretofore to the dreams of a few visionaries and the feeble efforts of a few clumsy inventors. We, ourselves, have thought that, in the present state of human knowledge, it contained no promise of success. But, considering the greatness of the prize and the trifling character of the endeavors which have been put forth to obtain it, would it not indeed be well, as our correspondents suggest, to make a new and combined effort to realize it, under all the light and power of modern science and mechanism?

What little attention this subject has heretofore received from inventors has been almost wholly confined to two directions — flying by muscular power and the guidance of balloons. Both of these we have been accustomed to regard as impracticable. But, as Mr. Hyatt suggests, the flying by muscular power is a field of invention which has by no means been thoroughly explored. Though it may be impossible for a man to raise his own weight rapidly by beating the air, the *sustaining* of his weight in the air and moving horizontally is an entirely different problem. In the bird the wings are moved by the most powerful muscles in the system. Has this hint been acted upon and the muscles of the legs and shoulders been brought to bear upon the wings in the most efficient manner? Again, has the constancy of the rotary motion been made available in a flying machine? If spiral fans were used, of course, two sets would be required to prevent the machine from turning itself in the direction opposite to the motion of the fans.

But the thing that is really wanted is a machine driven by some natural power, so that the flyer may ride at his ease. For this purpose we must have a new gas, electric, or chemical engine. What we require are two or more substances, solid or liquid, which, by merely being brought in contact, would be converted into gas. Place these in the reaction or Avery engine, which, by running at high velocity, would yield a large power in proportion to its weight, and it is possible — yes, probable — that the machine would drive spiral fans with sufficient force to raise itself from the ground. Would not the binoxyd of hydrogen and charcoal fill these conditions? This engine would run with such immense velocity that the fans would have to be very small, and it is probable that a moderate widening of the arms themselves — giving them a spiral inclination — would be the true plan. There might be two generating vessels, corresponding to the steam boiler, and when one was exhausted, the second might be brought into action while the supply of material was renewed in the first, thus supplying and exhausting them alternately.

The simplest, however, of all conceivable flying machines would be a cylinder blowing out gas in the rear and driving itself along on the principle of the rocket. Carbonic acid may be liquefied, and, at a temperature of 150 degrees, it exerts a pressure of 1,496 pounds to the square inch. If, consequently, a cylinder were filled with this liquid and an opening an inch square made in the lower end, the cylinder would be driven upward with a force equal to 1,496 pounds, which would carry a man, with a surplus of some 1,350 pounds for the weight of the machine.

We might add several other hints to inventors who desire to enter on this enticing field, but we will conclude with only one more. The newly discovered metal aluminum, from its extraordinary combination of lightness and strength, is the proper material for flying machines.

Like Saint-Exupéry's, this piece is also about the early days of airmail—on the other side of the Atlantic, in the winter of 1934, and in some of the worst weather imaginable. Survival in that perilous time was dependent not only on skill but to a perhaps greater degree on luck. Or call it divine will, as Colonel Robert L. Scott, Jr., did in the title of the 1943 book excerpted below, *God Is My Co-Pilot.*

ROBERT L. SCOTT, JR.

When Death Flew the Mails

If you remember 1934—there was trouble between the government and the airlines concerning airmail contracts. To me even this was a lifesaver in securing flying time, for all of us had recently been ordered to fly no more than four hours a month. This was the bare minimum to receive flying pay, and, as it turned out for many, the best way to get killed in airplanes. It's still a game that takes constant practice.

The first incident to permit us to keep flying temporarily happened when the youngest pilots at the station were sent to Chapman Field near Miami for gunnery camp. We left the snows of the East and went to the sunshine of Florida, about February 1. From this camp we were saved again from going back to Mitchel Field and to the rule of no more than four hours a month. On February 13 came orders to take off for Cleveland for the job of flying mail over the Alleghenies from Chicago to New York. I completed loading my Curtiss Falcon, got my crew chief, Sergeant Tetu, aboard, went to the down-wind end of the single runway, and gave the ship the gun.

We took off and as we pulled up over the big Bellanca transport at the other end of the field I felt something peculiar about my ship. I began to climb slowly to the left—and then I noticed a blast of wind across my face. I was skidding. I tried to correct it with the rudder and it became worse instead of better. Exerting more pressure on the pedals, I looked down, and I could see the springs of the rudder control moving. But the rudder itself was not moving. It was stuck. I looked back over my shoulder with a peculiar feeling at my heart and saw the rudder locked in the full left position.

I didn't know what to do. I pushed the nose down, and as the speed increased the skid got worse. I turned to Sergeant Tetu and started to yell to him to bail out, and then I thought: We're too low for that. So I just began to fight the ship. I would give it full gun and then slide the throttle gradually back to see if by crossing the controls (by keeping the stick forward and to the right) while the rudder was locked to the left, I could keep the ship in a mild slip but still have control. I don't know whether I figured this out or just luckily did it — for things happened too fast.

We circled the field and came around, while I tried to land on the runway into the wind, but as I went to cut the gun the ship nearly spun in and I almost hit the transport as I dove out to recover. I gave it full gun again and climbed, with my right leg braced against the rudder trying to bring it free. I then thought: "This is a time for a crosswind landing if there ever was such a time — because the wind is pretty strong and if I come around with full left rudder on and can push the stick far over to the right, it will be the same as slipping toward the ground, and if I'm slipping toward a 20- or 30-mile wind it may hold us straight till we get the wheels on the ground."

At any rate, I did this and we scraped the trees coming over the mangrove swamp near the field boundary. I slipped and skidded toward the hangar and the ship hit the ground. I cut the switch and climbed out, and I assure you that the first thing I did was to reach down and pat the ground lovingly.

Then I looked at Sergeant Tetu, who was a kind of greenish color. "Lieutenant," he said, "I thought you were just showing off at first. When you dove down on that transport I thought you were giving them a scare, but when you did it the second time I knew there was something wrong." We unzipped the fabric inspection plates from around the tail, and in the empennage section we found about a bushel of empty machine-gun cartridges, one of which had jammed in the rudder pulley and had locked the rudder in full left position.

We proceeded on up into the cold of the North, landing in Cleveland, Ohio, that afternoon. And then began the greatest training that I have had. I wasn't going to training school this time: I was in something more important. Here was what we had been prepared to do — or unprepared to do. Here we were, about to start out flying the mail in tactical planes with open cockpits, in the blizzards of the Great Lakes region, the Rockies, the Northwest, in the cold of the prairies. Would they work or not? We certainly didn't know.

The weather we flew in to carry the mail during the winter of 1934 was about the worst in history. I sometimes think the powers on high collaborated to give us a supreme test. There were fourteen pilots killed along that airmail run, and most of them were killed because we had no instruments for the ships, or at least not the proper type for flying blind. We flew pursuit ships, which carried 55 pounds of mail; we flew old B-6 bombers that would carry a ton of mail at a speed of 80 miles an hour, providing the wind in front of you wasn't too strong—sometimes they almost went backward. We flew everything from a Curtiss Condor, which Mrs. Roosevelt had been using, to the old trimotored Fords. And we flew through the worst weather in the country.

The route that I flew from Chicago, to Cleveland, to Newark, was what was known to all airmail pilots as the "Hell Stretch"—and it was just that, as I found out pretty quickly.

We stepped out of routine flying into something we knew nothing about. I took ships that I had never seen before and flew them on orientation flights toward Chicago. I had never been to Chicago. One of these ships—a P-12-K with a fuel-injector system—I flew West out of Cleveland one morning, but I didn't know anything about changing the fuel tanks. You see, usually you just let the gasoline run out of one tank and then you turn another on before the engine quits running from lack of fuel. But in this type of injector system, if you should let the fuel run all the way out of the tank, the engine wouldn't pick up when you turned on the selector valve!

Over Michigan City, Indiana, running out, I reached down to turn on the other tank. The prop just windmilled and I had to land down on a broad, open, snowy plain. When I landed, I found it was the airport of Michigan City, which had been hidden by three feet of snow. I opened the technical pamphlet carried in the ship, read in the book how to work the fuel system, turned the tank on there, and using a wobble pump got the fuel back up to the engine. But then I found out that there was no crank for this type of ship—you had to use a shotgun shell to start the engine. I was learning things every day.

Well, I crawled around and finally after reading everything again, I got the engine started by pulling a trigger and exploding the shotgun shell. And after losing about three hours, I bounced across the snowdrifts and went on to Chicago. I figured that everyone would be standing out on the field, worrying about what had happened to me and the P-12-K. Landing, I ran hurriedly in to the flight officer and said, "I'm the pilot who came here from Cleveland, I had to land in Michigan City . . ." and so forth. One of them said, "Well, we didn't even know you were on the way—our radios aren't working and the PX systems haven't been installed by Western Union." That began to show me that things weren't so well organized.

And so we began our airmail flying—slightly SNAFU, as we have learned to say from the gremlins in this war. There were tragedies and there were funny things. Among the tragedies, I remember one man had three roommates killed—so many killed that he refused to have anyone else live with him. Finally when my second one was killed, we lived together—figuring rightly that better things were coming.

On February 19, 1934, we made our first real airmail runs from Cleveland, some of them going to Newark, some down toward Cincinnati, some to Chicago. One of the funniest things I remember happened that night. We had got the first ship out, Lieutenant Bob Springer in a P-12-B for Cincinnati. We had worked over his ship to get him off on time, which was somewhere near two o'clock in the morning, out in the freezing snow of Cleveland airport. It must have been 13 below zero. Anyway, we got the ship started, saw Springer's light disappear in the dark, breathed a sigh of relief, and went into the hangar again to get another ship ready to go out. Just as we started to get settled down over the manifest for the next load West, we heard the roar of a P-12 come over and land. Outside we could barely make out the wing lights as it taxied back in the blizzard.

It was Bob Springer. He leaned out of his ship and yelled, "I can't see my compass — the light up there is out — get me a flashlight so I can find my way out of this town." We got him a flashlight. Even then, when he finally got away carrying something to read his compass with, one of his gloves blew off and he had to sit on the ungloved hand to keep it from freezing.

Sometimes people on new jobs got mixed up and sent the Cleveland mail in the wrong direction from Chicago, toward Omaha, or sent the Chicago mail from Cleveland to New York, the reverse direction — just normal events amid the "growing pains" of an army flying the mail.

Once the control officer finally got a man in the air after sweating the weather out to the West for days. I saw his ship take off and disappear in the snowstorm. Then I saw Sam Harris jump up, for the U.S. mail truck had just driven up. It was late, and in the excitement of getting the ship's clearance the eager pilot had forgotten to wait to have the mail loaded. The control officer had to call him back and start all over.

Buster Coln, a former college boxer from Clemson, used to give the newspapers some wild stories of the trials of bad weather flying. I'd say to him, "Buster, you told the papers you landed last night with about a pint of gas — just enough to get you over the fence. Well, I looked at the record and you had half a tank." Buster would look up and say, "Hell, you got to give these paper boys something good — they wait for us out here all night and I'm not going to land and tell them everything was routine."

His most exciting yarn grew out of a flight when he was really about to give out of fuel over North Platte, Nebraska, in a snowstorm. Buster told it like this: "I decided to jump. So I climbed the A-12 to 6,000, got it straight and level. Then, after getting out on the wing — still holding the stick — I said to myself: 'Buster, you don't want to do this and leave this fine attack ship to crash from up here.' So I got back in, buckled my safety belt, and tried again to fly down low enough in the blizzard to see where I was. Finally I saw a light and dropped a flare — I cut the engine and glided down to try to land. I hit the ground pretty easy without even seeing it, dodged trees and snowdrifts, and finally came to a stop.

"A farmer came out and took me and the mail to the train, and I was taken to the home of the mayor of North Platte, where there developed quite a party which lasted until the storm was gone. We then went out and dug the ship out from under the snowdrifts. I had landed on the levee of the Platte River and had come within inches of hitting trees — so I guess the Lord was with me. The mayor had the WPA boys in the city pull the plane to the best place for takeoff, and then after getting it serviced and the top of the levee scraped smooth with a snowplow, I climbed in to go on to Omaha.

"When I finally got the engine started after wearing out most of the WPA boys on the crank, I looked down the improvised runway and contemplated the takeoff. The mayor ran over about that time, and holding his hat on his head in the slipstream, he yelled, 'Can you get over that tree down there, Lieutenant?' I said I didn't know, so the mayor waved his hand at the WPA men and yelled: 'Cut it down.' I climbed out and waited. When this obstacle had been removed, I got back in, but the mayor pointed toward an old barn far down the runway and asked if I could get over that. At my second reply of 'I don't know,' he called to the men and said, 'Burn it down.' So, I had to stay with him that night for another party. Next day I got the ship started again and finally got away."

The mayor of North Platte was quite a boy, and so was Buster Coln. His best stories used to rival those of airmail pilots on the West Coast, and the best of them all was good enough for the Liars' Club—which is the classic of the air corps.

A pilot out there was having trouble over some very thick weather. He tried several times to land in a tule fog in the San Joaquin Valley, on the Fresno Airport. Each time he'd come closer and closer to the ground, and then, seeing nothing, he'd have to go around again. Finally the plane touched the ground, light as a feather, and without rolling more than 10 feet it stopped. The tired pilot looked all around but could see nothing. He climbed out and with his hands outstretched to keep him from walking into something, he took three slow steps and felt his hands touch a hard brick wall. He walked forward, holding one hand on his plane's wing, stretched out his other arm, and found another wall. Continuing all around the ship he found to his surprise that there was a wall all about him. Next morning the tired pilot, having slept in desperation under the wing of his plane, found that he had landed in a silo.

Physicist, writer, lecturer, and public servant, Charles Percy Snow, Baron of Leicester, was profoundly concerned about the effects upon society of advancing technology. This article, commissioned by *Look* in the days following the moon landing of July 20, 1969, contains such disquieting observations as: "At present, we are letting technology ride us as though we had no judgment of our own," and ". . . will the landing have an effect on us? . . . I am afraid it will." If God's heaven failed to retain its hold on our imaginations, Lord Snow asks, how long will space maintain its allure?

C. P. SNOW

Moon Landing

We have seen a wonder. There has never been one quite like it. What first steps in human history would one have chosen to witness, if one could travel in time? The Vikings coming ashore wherever they did come ashore — Newfoundland? — in North America? Or the first little boat from Columbus's ship scraping the land under her keel? Yet all of that, or any other bit of geographical discovery, we should be seeing with hindsight. On the spot, it must have seemed much more down-to-earth. People getting out of boats must have looked (and felt) very much like people getting out of boats anywhere at any time.

No, we have had the best of it. We have seen something unique. It is right that it should have looked like something we have never seen before. In science films, perhaps — but *this was real*. The figure, moving so laboriously, as though it was learning, minute by minute, to walk, was a man of our own kind. Inside that gear there was a foot, a human foot. Watch. It has come, probing its way down — near to something solid. One expects to hear (there is no air, one could hear nothing) a sound. At last, it has come down. Onto a surface. Onto the surface of the moon.

Well, we have seen a wonder. We ought to count our blessings.

This is the time when we might try to clear our minds about the whole project. There will never be a better time. If the landing had failed, we shouldn't have been in a mood to be even moderately detached. But now we are happy, admiring, and basking in a kind of reflected moonlight. If we can't ask sensible questions now, we never shall.

It is important to ask the right questions, though. For instance, I don't think it's realistic to wonder whether a moon landing should ever have been attempted. It may have been wasteful; it may have taken up energies and resources that neither of the superpowers could comfortably afford. We will consider that in a moment. But once the project was technically feasible at all—and no one has doubted that for a good many years—then *it was bound to be attempted.* Not for any of the highfalutin reasons that we have all bandied about.

Some of them existed, but they were rationalizations to make the decision seem more praiseworthy. Just as when people wanted to climb Everest or trek on foot to the South Pole, there was a lot of talk about the scientific results to be obtained; the great polar expeditions before World War I got financed in the name of science. That was non-sense. The scientific achievements were minimal. The scientific results of the moon landing aren't minimal, but they wouldn't justify the project, and they weren't the ultimate reason for it. Nor were the military possibilities, though credulous persons in both the United States and the Soviet Union presumably at times thought they were. That bogey can be disposed of later.

National prestige was more like a genuine reason, but it didn't have to be proved to a reflective outsider that both the superpowers—once they became committed to space exploration—would emerge approximately equal, sometimes one leapfrogging the other, and vice versa. For technology doesn't speak either English or Russian, and—in a competitive activity, war or space or anything you like—no technological lead is ever held for long.

No, the project was bound to be attempted for a much deeper or more primitive reason. If there is a chance for men to put their feet where other feet have never been, they will try it. The answer of George Mallory, the English mountaineer, has become a tedious cliché, but like a lot of clichés, it is true. Why do you want to climb Everest? Because it is there.

Well, the moon was there. The moment there was a chance to get on it, someone was bound to try. The process was accelerated by national rivalries, but it would have happened even if the United States or the Soviet Union alone had had a monopoly of rocketry. For any great country has a supply of brave and spirited men who would have been ready for any adventure technology might give them. That is grand; it makes one proud of belonging to the same species. But there is something else that is perhaps not so grand, that is unarguable and also sinister. That is—there is no known example in which technology has been stopped being pushed to the limit. Technology has its own inner dynamic. When it was possible that technology could bring off a moon landing, then it was certain that sooner or later, the landing would be brought off. However much it cost in human lives, dollars, rubles, social effort.

I do not find this reflection consoling. So far, the space project has not done practical harm. In money, it has cost a great deal, as well as in social effort. In human lives, so far as we know for certain, it has cost just four, three American and one Russian — a good many less than were lost before Everest was climbed (Mallory himself died there) or the South Pole expeditions got home. That is something to be thankful for. But it would be reassuring to find one case in which technology was called off: when human sense and will said, yes, that feat could certainly be done, but it isn't worth doing. When that happens, it will be a sign that we are gaining some hold on our destiny. At present, we are letting technology ride us as though we had no judgment of our own. There were a good many possible justifications for space exploration, and this wouldn't have been a good place to stop technology in its tracks. . . .

But one can't — or at least I can't — say the same for the faster-than-sound passenger aircraft. Up to the present moment, billions of dollars have been spent by the United Kingdom, France, the Soviet Union, the United States, developing these machines. America can probably afford it. Britain can't. The Anglo-French *Concorde* is an extraordinary example of conspicuous waste. The time-saving in air journeys will be trivial, especially as the ground traffic to airports gets worse. Yet here again, the demon of progress in technology is making us all behave as though we had our brains amputated.

Of course, the moon landing and space exploration in general have cost the United States and Soviet Union prodigious amounts. There have been times when I might have said — and in fact did say once or twice — that the money could have been better spent. But now I have had second thoughts, and I believe that I was wrong. I was making what I said at the beginning was the mistake we have to avoid, that is of asking the wrong question. If one asks, Could the money spent on space exploration have been used for something more valuable? then the answer is, *Of course*. But it is not the right question. The real question is, Would the money, if diverted from space exploration, have actually been used for something more valuable? And there the answer is, Almost certainly not.

Imagine that big rockets had proved unworkable, so that we couldn't project ourselves outside the earth's gravitational field. In that case, NASA wouldn't have existed; and in theory, a gigantic sum of money would have been free for other purposes. Does anyone really believe that most of it, or even a good share of it, would have gone to worthy, down-to-earth purposes, like making cities inhabitable once more? If anyone does believe that, he doesn't know how politics in our kind of society works. It is equally implausible to imagine that the money would have been devoted to increasing the food supply, and decreasing the population growth, in the poor countries of the world. Ask Robert S. McNamara, who, as president of the World Bank, is living with this nightmare problem, whether that could conceivably have happened. The only cause that might have seized the American imagination on a big scale is cancer research. A little of the investment might have been diverted in that direction. But, curiously enough, it would be quite impossible to spend the vast NASA funds on any pure or applied research in molecular biology or some other area immediately valuable to man. Even for nuclear physics, the most expensive of researches, there is, by the standard of military or space budgets, a fairly low saturation limit. It is singularly difficult to spend really big sums on entirely benevolent practical projects: either the projects can't absorb the money, or they have no political sex appeal. It isn't in the nature of politics for people to make themselves think ten years ahead — which is why we are running into a population-food crisis without sacrificing one night's sleep. The space project is probably — apart from military things — the only spender of big money that could have collected majority support from politicians and public. So we are misleading ourselves — and flattering ourselves — if we call it a waste.

Much the same seems to have been true in the Soviet Union. The budgeting is performed in a different fashion from ours: it is more completely centralized, and is of course not argued about in public. Nevertheless, similar considerations apply. The Soviet Academy of Sciences is allotted a massive block grant that it divides up among research projects — including many that we should think of as development. (This is as though the National Academy of Sciences in Washington was given funds to finance all research in all American universities and elsewhere.) But the space expenditure, like the military expenditure, must have been decided at a still higher level — one would guess in the Politburo itself. No doubt some of the Soviet academicians, like their American opposite numbers, complained that the money could have been used more sensibly, in particular on their own pet subjects. But their complaints got nowhere and couldn't have got anywhere in America either.

There is one theoretical difference. If the space money in both countries had suddenly become available for other purposes, it is conceivable that the Soviet bosses would have found the problem easier to handle. Concentration on specific large-scale projects is something they are well adapted for — which, incidentally, is why they have done so well in space. But whether they could have used their equivalent of NASA funds on, say, housing or the rapid development of the Siberian mines is very doubtful. Politics systems differ, but the difficulties of going faster than natural cussedness permits stay much the same.

The question of waste of money and effort becomes more misty the longer one looks. So does the military significance of space, though no doubt influential persons on both sides took it seriously almost until the present day. It is characteristic of the human tendency to become bedazzled by gadgets. Influential persons think, one more gadget, one more bright idea or piece of hardware, then we have an overwhelming military advantage. *Very fortunately*, the world is not so unstable as that: if it were, we should all be fastening our seat belts into futurity every hour of the day. The world is not so unstable, and even modern technological war is a good deal simpler, and looks as though it will remain so for the rest of the century. Once both sides had sent a rocket round the moon — and probably, to those who kept their heads, long before that, as far back as the first satellites — it was all exaggeratedly simple, as simple as the Mosaic laws. If you can send a rocket round the moon, you can take out Moscow or, alternatively, New York. You can take out (I apologize for using this inhuman jargon) any town anywhere on earth. Nothing can stop you. It is as simple as that. Simpler than neolithic warfare used to be. That is the balance under which we live. *Very fortunately*, it is a balance difficult to perturb. Whatever afflictions we run into, and there may be many, thermonuclear war between the superpowers is perhaps the most unlikely.

In the face of this cataclysmic simplicity, it is rather odd that people get excited about space landing platforms and so on. It is perfectly true that they could be used for discharging missiles—but that could be done much more conveniently from earth. It is also perfectly true that they could be used as sources of surveillance or spying; so could existing satellites at disproportionately less cost. Further, there are five or six different methods of getting military information, much more efficient and much cheaper; the U.S. and the Soviet are thoughtfully using them upon each other every day of the week. But responses to spying aren't at all rational; people blissfully ignore the really effective methods and then get feverish about the James Bonds and Le Carré's heroes, who could disappear without making one cent's worth of difference.

People also get feverish, and feel a sort of superstitious horror into the bargain, at the thought of being watched from above. It doesn't merit all that fuss. No position in space is going to affect the balance of deterrence by one percent.

In fact, the competition in space between the U.S. and the Soviet Union has done no harm, or singularly little. There has been a difference of emphasis: The United States in the Kennedy administration set its sights upon the moon, and the Soviet Union has been thinking in addition about planetary trips. But neither side would have been impelled to go so fast if the other hadn't existed. That speed has certainly used a lot of money and a lot of very rare skills, in which there would have been economies if the two countries, instead of competing, had cooperated. But, granted the history in which we have lived, it seems inevitable that the two sides would have competed somewhere. If not in space, then somewhere else. Quite possibly, more probably than possibly, somewhere more harmful. Granted that there was going to be a competition, it would have been moderately difficult to invent one more innocent. (Imagine that the two countries had settled on association football as the chief test of national prestige: the expenditure would have been less, but, to judge from international experience, the loss in temper, rancor, and even life would have been significantly greater.)

Most of the questions about the moon landing (taken to be the symbol and emblem of the entire space project) don't seem, then, to be very realistic. They are the wrong questions. But there are a few questions really worth asking. One is, have we gained anything, and, if so, what? And what effect is the landing going to have on us? Will it change our lives? How, and to what extent?

As for the first, yes, we have gained something. The most important part of it is a moral gain—if men do something wonderful, then it ought to be, and often is, a source of hope for the rest of us. We are compelled to spend so much time in this jagged century looking at the worst in other people and ourselves. We should be stupid and guilty if we didn't. But it is just as well that we should have the occasional spectacle forced upon us of men being clever, competent and brave.

Human beings are imitative animals. They imitate crime—not only the addiction to crime, but the way it is performed — with the utmost enthusiasm. It is a blessing when we are given something else to imitate: the technological and organizational skill of the whole NASA team, as well as the cool professional courage of Neil Armstrong and his colleagues.

There is also a clear scientific gain. We now know beyond doubt a good deal about the constitution of the moon; soon, we shall know more. This means that before long, we shall solve some tough scientific problems about how the solar system was formed. Further, we shall before long be able to erect laboratories on the moon or a space platform. That doesn't sound dramatic, but the result may be very dramatic. At the moment, cosmogonists are arguing and guessing about the nature of the universe. Observation outside the earth's atmosphere should tell us some of the answers. So the immediate human gain is large, and the scientific gain considerable. So far, so good. But will the landing have an effect on us? Will it change our lives?

I am afraid it will. I am afraid that in the long run, perhaps a generation, perhaps longer, it will have a bad effect. It will give us the feeling, and the perfectly justified feeling, that our world has finally closed in. This is forever the end of the mortal frontier.

I dislike saying what I have just said, and am going on to say. No one is fond of stating a negative opinion. Too many such opinions, even by men we consider tremendously wise, have turned out to be wrong. On that account alone, no responsible person would put forward a flat negative opinion unless he was unusually convinced.

Further, it is no fun putting forward any sort of opinion, positive or negative, unless you can see it checked one way or the other. This one can't be, in my lifetime and for far longer, maybe for hundreds of years. Yet I feel it would be cowardly not to speak my mind. There are some things on which we deceive ourselves very easily. Science presents us with many horizons to which we can't see the end. Fine. That doesn't mean that all horizons are infinite. I am sure this one isn't. (Yes, I know all about tachyons, those hypothetical particles moving faster than the speed of light.) The horizon is limited because of the size of the universe and the shortness of a human lifetime.

This is the only point on which I flatly disagree with the space enthusiasts. They speak as though reaching the moon (and the other possible spots in the solar system) is going to liberate the human imagination as the discovery of America did. I believe the exact opposite, that the human imagination is going to be restricted — as to an extent it was when the last spots on the globe had been visited, the South Pole and the summit of Everest. Nowhere on earth for adventurous man to go. Very soon, there will be nowhere in the universe for adventurous man to go.

The analogy with the discovery of America is a very bad one. The Spaniards and Portuguese found riches, marvels, above all, people, in the American land. When Armstrong trod on the moon, he found lumps of inorganic matter, as he might have done at the South Pole. The South Pole is a pretty accurate analogy.

John Donne, writing his poem in the early seventeenth century, when the discovery of America was still fresh in English minds, rhapsodized about his mistress as, "O my America! my new-found land." Can anyone imagine him comparing the girl to the South Pole or the bleak and desert moon?

The trouble is, the solar system is a desperately disappointing place. Scientists have known this for a long time; it is now being confirmed in concrete, only too concrete, fact. Our planet is a peculiar fluke in a dead system. Before the end of the century, Americans and Russians will have landed on Mars. It is conceivable that they will discover traces of primitive organic life. Scientifically, that will be exciting. But that is the very most that we can anticipate. A little lichen on a barren world.

Where else can we go? One can tick off the possibilities on the fingers. One or two of the minor planets, perhaps. Just imaginably, but only just, one or two of the moons of Jupiter. Those, we can predict, will be more barren lumps of inorganic matter. Then we come to the end. That is the frontier. There is nowhere else in the entire universe where man can ever land, for so long as the human species lasts.

This has been scientifically obvious for long enough. The solar system is dead, apart from our world; and the distances to any other system are so gigantic that it would take the entire history of mankind from paleolithic man to the present day to traverse — at the speed of *Apollo 11* — the distance to the nearest star. So that the frontier is closed. We can explore a few lumps in our system, and that is the end. This has been, as I say, scientifically obvious for a good many years. But it sometimes takes a long time for scientific certainties to reach the public, even the educated public (think of the singular ignorance about human genetics). It takes the evidence of eyes and hands and feet — the eyes, hands, and feet of Armstrong and his successors — to translate the scientific certainties into common knowledge. In the domain of space travel, I reckon that this will take from thirty to one hundred years. Then disappointment, the curse of confinement, a kind of cosmic claustrophobia will set in.

We had better be prepared for it. It is no use holding out the prospect of limitless horizons when the horizons are certain to turn out only too desolately limited. Science fiction writers (at least those who have scientific knowledge and insight, of whom there are plenty) have been fighting a rearguard action against the inevitable. But they will get tired of postulating relativistic biology and relativistic speeds. In fact, one of the casualties of the moon landing will be science fiction, at least as applied to space travel. You can write scientifically about what you know to be improbable, but you can't write scientifically for long about what you know to be impossible. Science fiction writers will be driven inward, not outward, and will turn their attention to human biology and psychology. There is plenty to occupy them there for more than a hundred years.

Nevertheless, they will be driven inward. So will the imaginations of the rest of us. It is a paradox. The greatest exploration — that pioneers have been looking forward to so long. The greatest exploration. The pioneers weren't to know the realization that would afterward gradually dawn upon us. The realization that, as well as being the greatest exploration, it was very near the final one.

That is a pity. We badly need something to take us out — not constantly but for part of our time — out of this, our mundane life. The naive idea of heaven did that for generation after generation, when people could believe that heaven was somewhere above us, up there beyond the sky. It was a naive idea but very powerful, and nothing more sophisticated has been so powerful. To secular minds the prospect of space, other worlds to find, other lives to meet, has been a substitute. To many a substitute of almost equal power. Now that will fail us, too. As a result of supreme technological skill and heroism, we are faced not with the infinite but with the immovable limits. The limits of our practical condition. We now know that the only lives we shall ever meet turn out to be our own.

Those who saw him fly say there never was his equal, and never will be. Lincoln Beachey buzzed Michigan Avenue in Chicago only eighteen inches above the traffic. He flew low around a racetrack, flicking up the dust with his wing tips on the turns. He zipped between trees so close together that he had to bank 60 degrees to fit between them. And he invented the Vertical Drop, the Dutch Roll, the Turkey Trot, and other daredevil stunts. He met his end in a Taube monoplane, an 80-horsepower speed demon of a craft whose flying weight—including aviator and fuel—was only 600 pounds. To prevent being tossed out during a wrenching maneuver, Beachey had himself strapped into the open cockpit at the waist, legs, and ankles; it was this uncharacteristic safety precaution, not his plunge into the bay, that killed him.

RICHARD F. SNOW

Lincoln Beachey

"They call me the Master Birdman," he said once, "but they pay to see me die." He hated his audience. He loved his audience. He was bitter, contradictory, expansive, and fatalistic. Foremost in the first generation of daredevil pilots, he flew in a natty pinstriped suit with a two-carat diamond stickpin keeping his necktie in place. His fellow pilot Beckwith Havens, one of his very few close friends, described him as "a strange, strange man." He was also, according to Orville Wright, who knew something about it, "the most wonderful flyer I ever saw and the greatest aviator of all." He claimed to have flown for twenty million people—and he flew wide enough and far enough for that to be possible.

Lincoln Beachey was born in San Francisco in 1887. While still in his teens, he made his way east to Toledo, where he got a job in the balloon factory of a man named Charles Strobel. He wanted to fly, but Strobel refused to let him, so Beachey started spending his nights in the factory, sneaking out at dawn to take up the airships. After a few weeks of ghosting around over the sleeping town, he told Strobel what he had been doing, and demanded an aviator's contract. He got it.

By 1905 he was the best in the business, and sometime in 1910 he decided to shift from airships to airplanes. It was a typically reckless Beachey decision; more than thirty pilots died that year trying to put their wood, wire, and canvas machines through paces beyond their capabilities. Impressed by his reputation as a balloonist, the Wright brothers offered to take him on as an exhibition flyer, but the money wasn't good enough, so he went to Hammondsport, New York, where a gifted inventor named Glenn Curtiss was building airplanes. Curtiss gave him a tryout, but Beachey immediately wrecked the plane. "His big trouble," his brother Hillery said, "was that he wanted to stick it right up into the air. There wasn't enough power to do it. . . . He broke up several of Curtiss's planes. . . . Curtiss was afraid to look—just turned away when he first saw him fly." But Curtiss's exhibition manager saw something in Beachey's flailings and browbeat Curtiss into sticking with the impetuous would-be pilot. It turned out to be the best publicity investment Curtiss ever made, for as soon as Beachey got the hang of handling an airplane, he flew like a drunken angel.

In an era when most people were awed just by the sight of a plane in the air, Beachey could make his primitive machine do almost anything. Bucking and twisting across a field, he would angle down to pick up a handkerchief off the ground with his wing tip. Then he would climb a mile up, cut his engine, and go into his "death dip," a vertical dive that had women in the crowd fainting.

Beachey's first summer's record is an indication of his ability. In June of 1911, with 150,000 people watching, he dove into the gorge of Niagara Falls, came out through the spume at the base, and flew under the International Bridge. With his carburetor sucking spray and his engine failing, he barely managed to scramble up from the boiling rapids to the safety of the Canadian shore. He never tried that particular stunt again, and neither has anybody else.

A month later he picked up a five-thousand-dollar prize for flying from New York to Philadelphia. Two weeks after that, he set the world's altitude record by topping his fuel tank and then simply climbing as fast as he could until, in the arctic air currents more than two miles up, his gas ran out.

If he celebrated after he glided back to earth, it wasn't with liquor. "One glass of champagne, and he'd be tight," Havens recalled, surprised by Beachey's abstemiousness in an age when at least one of his fellow pilots drank so hard his mechanics had to lift him into his plane. Beachey's failing, "a real strong weakness" according to Havens, was women; he left a string of disappointed "fiancées" behind him as he barnstormed around the country.

Suddenly, at the height of his fame in 1912, Beachey announced his retirement. "I have defied death at every opportunity for the last two years," he said. "I have been a bad influence, and the death of a number of young aviators in this country can be traced, I believe, to a desire to emulate . . . my foolishly daring exploits. . . . You couldn't get me in an airplane again at the point of a revolver."

This moody resolution was short-lived. Beachey had spent a few miserable months on the vaudeville stage when he got word that a Frenchman had looped the loop. "If he can do it, so can I!" Beachey shouted, and told Curtiss to build him a special stunt plane. Soon he was barnstorming again, charging $500 for his first loop and $200 for

Lincoln BEACHEY

ALTHOUGH HE WAS THE GREATEST AND MOST DARING OF THE EARLY BARNSTORMING STUNT PILOTS, BEACHEY WAS OBSESSED WITH AIR SAFETY. HE WANTED TO MAKE FLYING AS SAFE AS WALKING.

IN 1910, BEACHEY SOLVED THE SPIN MENACE, WHICH WAS BECOMING A COMMON AND NEARLY ALWAYS FATAL HAPPENING IN THE AIR.

LOGIC SAID TO FIGHT A SPIN, AND PILOT AFTER PILOT DIED TRYING.

BEACHEY REASONED HE COULD BEAT THE SPIN BY TURNING INTO IT. 12 TIMES HE DELIBERATELY WENT INTO SPINS AND 12 TIMES HE PULLED OUT SAFELY.

each that followed. "That was . . . in the Middle West where William Jennings Bryan was talking," his brother remembered proudly. "I think they both got $1,000 a day, Bryan and Lincoln Beachey."

He had baited what he called the "scythe-wielder" more than most of his generation by the time he came to San Francisco for the Panama Pacific Exposition of 1915. "The old fellow and I are pals," he said, but he was obviously feeling strain, and occasionally behaving erratically. Once, when the governor of the state stepped forward to congratulate him after a flight, Beachey turned away, went back up, and, circling the field, stripped off his clothes. When he landed, he snarled to his mechanic, "Let's hear what he has to say about me now."

On March 14 he came out to his machine looking grave and troubled. His mechanics thought he had some sort of premonition of disaster, but 50,000 people had come to see him, and he went up. He looped over the bay, then lowered the plane's nose for the death dip. Aviators among the spectators said he was dropping at nearly 300 miles per hour when the wings tore away.

While the stunned crowd watched silently from the shore, a diver from the battleship *Oregon* located the plane, and thirty-five minutes later it was hoisted from the bay. They found Beachey tangled in the wreckage, his hands torn from his desperate struggle to free himself. America's greatest stunt pilot had drowned.

"The Flying Flapper of Freeport," they called the Long Island teenager Elinor Smith in the late 1920s; she hated the term, feeling it drew attention to her sex rather than her skill. She made her first solo flight in 1927 at the age of fifteen and only three years later was voted the top female pilot in the United States, having broken most of the women's endurance and altitude records. The adventure which first brought her to national attention occurred in 1928, when she flew under four New York City bridges to settle a bet. Mrs. Sullivan's reminiscence is one of many by old-time pilots in George Vecsey and George C. Dade's *Getting off the Ground.*

ELINOR SMITH SULLIVAN

Not a Woman Pilot, Just a Pilot

By 1928, flying under bridges was passé. New York was a sophisticated area, but this character from Iowa came to New York and decided to fly under the Hell's Gate Bridge. This was fifteen months after Lindbergh, but he felt this would help him somehow.

He was stupid. He got caught in a downdraft and busted up his Jenny and got picked up, I'm glad to say, by a garbage scow. He was making life miserable around the field, and Herb McCrory, the photographer for the *Daily News,* told him: "Even Ellie here could do it." Like I was the lowest branch on the tree.

[What could Ellie say? Exactly what she did say: "I could do it." McCrory was not content with prodding the teenager into challenging the bridges. He promoted a little wagering between the manufacturers of the Jenny and the manufacturers of her plane.]

So here's the Curtiss people betting against me, and here's the Waco people betting on me. There was five thousand dollars riding on it. I decided to do it on a Sunday. Mack called my father and said he was sorry he got me into it, but if I didn't want to do it . . .

It put me in a position where if I backed out, I was yellow. Father put it up to me: "If you feel it's important enough." I didn't, but others had put up the money. I told Mack, "If you're just going under the Hell's Gate Bridge, that won't do it. Let's do the four East River bridges. Once could be a lucky break, but four have different clearances."

I was still nervous. I had just got my license at sixteen, and I didn't want to get in trouble. All that week I hung by my heels from all those bridges, checking every angle out. But there was one thing I didn't notice: workmen on the first bridge.

On Sunday morning they left the ropes and blocks hanging down. That narrowed my clearance. As I headed downriver toward the Queensborough Bridge, my heart leaped. I was heading for the narrow span between Welfare Island and Long Island City, and I had to remember that going too fast would be a problem, but if I went slow I could ride out any problem.

Then I saw a white scarf on the bridge. I knew it was Dave Oliver of Paramount News, cranking out the film. Right away, I knew this was going to get me in trouble. All the newsreel crews were there. Fox. Pathé. Trans-Lux. Everybody had heard about it.

I got under the first bridge and kept heading downriver, under the Williamsburg Bridge, under the Manhattan Bridge. Then at the Brooklyn Bridge, there was a navy destroyer, right across my path. I had to go sideways, make a vertical bank, over the destroyer. Then I went by the Statue of Liberty. Boats were blowing their whistles at me. Everybody knew about it.

I got back to the field, Mack yanked me out of the plane. He was in seventh heaven. I said to Mack, "I want to see you. I'll lose my license." Then he told me what happened. That barnstormer from Iowa kept saying my father had Red Devereaux in the cockpit flying for me. That's why Mack had all the newsreel boys out. That night it was in every newsreel theater in Manhattan. "Teenager Flies the East River Bridges." It made the Chicago fire look like a Boy Scout picnic, believe me.

Three days later I got a summons from New York City. Jimmy Walker was the mayor then. My father went with me. I got all dressed up in a skirt, looking like any other teenage girl. We met Major William Deegan, Walker's assistant, a typical Irish politician, very personable. . . . He took us in to see the mayor.

Naturally, Jimmy Walker knew my father from show business. I don't think he had connected the name until that moment. Walker looked at me and said, "You mean, this is the girl who flew under the bridges? This is the aviator we have to chastise publicly?" I can still hear the major saying it, to this day. There were just the four of us in the room.

Walker said, "It's funny for a while." I told him I was worried. I didn't want to lose my license. This was ten days afterward, when we finally got to see him. Anyway, he announced he was suspending me—grounding me—for ten days. And he made it retroactive.

The Department of Commerce sent me a letter from Washington, to censure me. I was only a sixteen-year-old. To be censured was not so bad. I could understand their feelings. But the letter from the Department of Commerce included a handwritten note from the man's secretary. A separate note. She said she was glad to see a woman flying, and asked if I would sign my autograph and return it.

No, *The Spirit of St. Louis* was not the first plane to cross the Atlantic nonstop; it was simply the first to be flown solo, and the first to travel the New York–Paris route. Lindy had been preceded in 1919 by two Englishmen, John Alcock and Arthur Whitten Brown, who took off from New-foundland in their Vickers Vimy and, sixteen hours later, touched down in Ireland. Alcock and Brown became heroes to the world, were knighted by George V, and were handed a check by Winston Churchill for £13,000 in prize money put up by the *Daily Mail*. These pioneers' day in the sun was brief, alas, as Alcock died in a crash in France in the year of his glory and Brown never flew again.

LOWELL THOMAS and LOWELL THOMAS, JR.

Alcock and Brown Fly the Atlantic

After Blériot's conquest of the English Channel, it was inevitable that the next major challenge for aviators of all nations would be a nonstop flight across the Atlantic. At first, the coming of World War I interfered with this, although as early as 1913 Lord Northcliffe and the London *Daily Mail* had offered £10,000 for a first nonstop aerial crossing in either direction. The flight had to be made in less than seventy-two hours, with the takeoff and landing points somewhere in the United States, Canada, or New-foundland on one side, and the British Isles on the other.

When World War I ended in 1918, commercial flying was still in its infancy, and the public, generally speaking, had a suspicion of airplanes whose job, first and fore-most, had been to spread death and destruction throughout the four years of conflict. Thousands of wartime pilots found themselves with nothing to do, and many turned to other professions despite all the flying experience they had acquired during the war. Still, some were determined to pursue flying careers no matter what the odds, and their ambitions received quite a boost in the very month of the Armistice when Lord North-cliffe and the *Daily Mail* renewed that £10,000 prize offer for a first nonstop transatlantic flight.

By the spring of 1919, the island of Newfoundland off the coast of Canada had become the busiest aviation center in the world. Airplanes were being crated in by freighter from England, landing fields and hangars were under construction, and aviators, mechanics, and navigators were arriving on every steamer. The intense preparations for flights across the English Channel in 1909 had been leisurely compared with the feverish activity that now gripped Newfoundland, the staging area for a far more spectacular and dangerous race.

At least four teams of British aviators had arrived by the middle of May and more were expected with each passing day. Two private fliers, Harry Hawker, an Australian, and Kenneth MacKenzie-Grieve, had pitched their camp at Mount Pearl Farm where they were running their Sopwith biplane through a series of tests. Captain Fred P. Raynham and his navigator, C. W. F. Morgan, were preparing their Martinsyde biplane on a landing strip at nearby Quidi Vidi, while at Harbor Grace, a third team, headed by Vice Admiral Mark Kerr, was assembling a huge four-engine Handley-Page biplane. With the bases, located within a 100-mile radius of each other, there was an air of competition, and all this was heightened by a growing international corps of newsmen who were sending out daily dispatches on the flight preparations. Naturally the Newfoundlanders were watching all these developments with intense interest and amazement.

Soon the Americans were in the running too, although their plans did not call for a nonstop flight to the British Isles, and therefore they would be ineligible for the *Daily Mail* prize. The United States Navy had decided to demonstrate the range and power of its new Curtiss Flying Boats by sending three of these planes on a flight from Newfoundland to the Azores, thence to Portugal and on to Plymouth, England. Even if only one of them made it to the Azores, America could claim the honor of first flying the Atlantic, although it would be over a shorter distance than the route from Newfoundland to the British Isles.

The navy planes made their base at Trepassey, Newfoundland, and American warships were assigned areas all the way across the ocean to the Azores as standby vessels for refueling and rescue duties. Moving swiftly to get into the air ahead of the British, all three American planes, the NC-1, NC-3, and NC-4, took off from Trepassey late on the afternoon of May 16. First, the unlucky NC-1 was forced down in the Atlantic within 200 miles of the Azores and the crew was rescued. Then, the hapless NC-3 lost its bearings in a fog, came down at sea, and was missing for fifty-two hours. But it taxied through rough seas for 205 miles while riding out a gale and finally arrived safely in the Azores. Meanwhile, the NC-4 under Lieutenant Commander Albert C. Read with a crew of four, including pilots Walter Hinton and E. F. Stone, did manage to fly all the way to the Azores, covering the 1,380-mile distance in about fifteen hours. Then several days later, the NC-4 continued on to Portugal and arrived in England near the end of May, completing the entire journey from Newfoundland in about two weeks.

So the NC-4 became the first heavier-than-air craft to cross the Atlantic and its crew won much temporary fame for the achievement. However, this still left unclaimed the *Daily Mail*'s prize for a nonstop flight between North America and the British Isles in less than seventy-two hours. Moreover, flying the Atlantic without rescue ships deployed along the way remained a far bigger challenge in the eyes of a watching world.

On May 16, the day the American planes had taken off from Trepassey, the British aviators on Newfoundland were straining at the leash. And when the word came through that Commander Read and the NC-4 had reached the Azores, Hawker and Grieve made a final inspection of their Sopwith, then roared out across the Atlantic heading for Ireland. Less than an hour later, Raynham and Morgan, not to be left behind, sent their Martinsyde down the runway at Quidi Vidi, but failed to get off the ground. In their haste to follow their rivals, they tried to take off in a crosswind and their plane crashed into a ravine, injuring Morgan so seriously that he had to give up flying.

For the next several hours, Hawker and Grieve flew on toward Ireland, but loose solder in the radiator clogged the water-circulation system, forcing the plane down into the ocean a thousand miles east of Newfoundland. They were missing and given up for lost for the next seven days until the news came through that a Danish freighter had rescued them at sea, and a highly excited British public gave them a heroes' welcome when they reached London. Harry Hawker went on to become famous in aviation as one of the founders of Hawker Siddeley Aviation, Ltd.

Two British flying teams had failed in their attempts to cross the ocean while Admiral Kerr was still assembling his Handley-Page biplane at Harbor Grace, hopeful of getting away on his own flight by early June. But he was facing the threat of stiff competition from a pair of Royal Air Force fliers who had entered the race: their names—Captain John Alcock and Lieutenant Arthur Whitten-Brown. They had arrived at St. John's on May 13, but their airplane was stored in crates on board another vessel which wasn't due for another two weeks.

Who were these new entries? Some four months before, Arthur Whitten-Brown, a thirty-three-year-old World War I veteran born in Glasgow of American parents, had been looking for a job. He also had hopes of soon marrying Kathleen Kennedy, the beautiful red-haired daughter of a major in the British Ministry of Munitions. But the all-important question of a job had to be settled first.

During his early years, Brown attended school in England and, even though he possessed American citizenship, he had joined the British Army to fight the Germans. As a lieutenant, he led his men in battles at Ypres and along the Somme, then transferred to the Royal Flying Corps as an observer. Twice shot down, on the second time he was taken prisoner and spent fourteen months in a German POW camp before being repatriated.

During the war years Brown had made an intensive study of aerial navigation and after the Armistice he began making the rounds of aircraft firms that were showing an interest in transatlantic flying. One afternoon he visited the Vickers plant at Weybridge near London, and the superintendent, Maxwell Muller, listened quietly as Brown stated his qualifications. Then Muller asked:

"You are a navigator. But can you navigate a plane across the Atlantic?"

"Yes," Brown replied, apparently without misgivings.

"Then we have a job for you," Muller continued. "The *Daily Mail*, as you know, is offering a prize of £10,000 for the first plane to fly the Atlantic nonstop. We want our plane to be the first to do it, and Vickers doesn't care about the money. The men who fly the plane can have that. We have the pilot, and we have the plane, so come along and meet both."

A few moments later Brown was shaking hands with Captain John Alcock, a twenty-six-year-old British pilot who had come out of the war with seven enemy planes to his credit and a Distinguished Service Cross among his decorations. From the point of view of personality, the two were a study in contrasts, but they developed a liking for each other almost immediately. Brown, the Scottish-American, was slight of build, dark-haired, rather quiet and reserved. Alcock was sturdy, round-faced and blond, with a ready wit, a dry humor, and very British.

In fact, Alcock was a pioneer of British aviation, and had received his flying certificate at the age of twenty when he became an instructor in aerobatics. During the war he had flown bombing missions against the Turks, including one daring raid on Constantinople. Like Brown, he too had been taken prisoner after he was forced to ditch his plane in the Aegean off Gallipoli.

Together, Alcock and Brown looked over their transatlantic plane — a Vickers-Vimy biplane powered by two 350-hp Rolls-Royce engines. A converted bomber, it had been earmarked for raids on Berlin and had a cruising speed of about 90 miles per hour with a range of some 2,400 miles. Considerable work remained before it would be ready for the flight, but by early May final tests were completed and the disassembled plane was placed in crates for the voyage to Newfoundland. The two fliers, aware of the hazards of their mission, put their personal affairs in order and Arthur Whitten-Brown told an understanding Kathleen Kennedy their marriage would have to be postponed until after the flight.

When they arrived at St. John's, Alcock and Brown took up residence at the Cochrane Hotel, the bustling headquarters of transatlantic aviators during that hectic spring of 1919. With their plane still en route from England, they spent the next few days trying to find a suitable landing strip, a task that wasn't so easy. Newfoundland was in the midst of a real estate boom, brought on by all this sudden demand for aerodromes, and property owners were asking sky-high prices for land that had even the remotest appearance of being level. Every day for a week, Alcock and Brown drove for miles over seemingly endless bad roads in an effort to find a suitable field, but it was a desperate search. The aerodrome at Mount Pearl Farm, recently vacated by Hawker and Grieve, would have been perfect, but the owner wanted $3,000 rent until June 15 and $250 a day after that. Another likely area was found near Harbor Grace, but here Alcock was confronted with a demand for $25,000 when he approached the owner. The New-foundlanders obviously were determined to make hay while they had the chance.

The Vickers-Vimy arrived and the tedious job of assembly began at Quidi Vidi where Captain Raynham generously offered the use of his aerodrome facilities, even though the runway there would not be long enough to permit the Vimy to take off with a full load of gasoline. Still looking for a landing field, Alcock and Brown divided their time between the plane and a search of the countryside until one afternoon their luck took a new turn. The owner of a large meadow near a place called Monday's Pool made them a reasonable offer. They promptly leased the land, and hired a labor force of thirty men who brought along picks and shovels to level the hillocks and remove the rocks and boulders. The work was completed on June 9 and Alcock flew the Vimy over from Quidi Vidi with a light load of gasoline.

For the next four days, Alcock and Brown virtually lived with their meteorological officers, but each day the weather reports were unfavorable, with high winds, fog, and rain continuing out over the North Atlantic. Coupled with their irritation over the delay was a growing apprehension that Admiral Kerr and his Handley-Page crew were almost ready to take off. But regardless of the weather, Alcock and Brown finally decided to wait no longer.

Early on the morning of Saturday, June 14, they showed up at the aerodrome in their electrically heated flying suits and, together with their mechanics, they began filling the tanks with a full complement of 870 gallons of gasoline and 40 gallons of oil. Their personal luggage was stored in a compartment near the single open cockpit. For food they took aboard sandwiches, chocolate, malted milk, and a thermos of coffee, along with a few small bottles of brandy and ale. Fitted into the area around the cockpit was a mail bag with three hundred private letters, and also their mascots, "Twinkletoes" and "Lucky Jim," two stuffed black cats which Kathleen Kennedy had given them before they left England.

At 4 A.M., they were again told that weather conditions were something less than perfect.

"Strong westerly wind. Conditions otherwise fairly favorable."

A few hours later, a crosswind was blowing from the west and the two fliers decided to wait a little longer in the hope that it would die down. But the morning hours gave way to the afternoon with the wind still blowing as furiously as ever.

Convinced that he could get the plane off the ground without a crash, Alcock made his decision. First, he asked the Vickers manager on the field for permission to take off, then he motioned Brown on board. Standing together in the cockpit, both fliers turned to say good-bye to their mechanics and a crowd of Newfoundlanders who had come down from St. John's to see them off.

Navigator Brown announced:

"Our objective is the Irish coast. We shall aim at the center of our target."

And Alcock, the blithe spirit, added:

"Yes, we shall hang our hats on the aerials of Clifden Wireless Station as we go by. So long."

At 4:10 P.M., the chocks were removed from the wheels — and the mechanics, hanging on the wings and tail, "let go" as Alcock gave both engines full throttle and the Vimy started across the turf. The two fliers were seated side by side in the cockpit as they headed over the slightly inclined runway into the westerly wind.

The plane lurched and lumbered forward for 300 yards before the wheels left the ground, and it was none too soon. A line of hills and treetops lay ahead, and gale-force winds were bedeviling Alcock's efforts to gain altitude as the Vimy slowly climbed to 800 feet and over the trees. With his machine now well under control, Alcock headed over the fishing fleet in Concepcion Bay and then on beyond Signal Hill to the open Atlantic.

It was at St. John's in 1901 that Signor Guglielmo Marconi received the first wireless signal ever sent across the Atlantic. And now, eighteen years later, the Marconi Station was flashing the news of the takeoff to every ship in the North Atlantic, asking them to be on the lookout and give the plane's position if sighted. Brown would be counting on this information to aid the navigation, and the moment the Vimy left the coastline he unstrung the wireless aerial to tap a message to the Marconi operator:

"All well and started."

Gradually, Alcock nosed the Vimy upward until they were at 3,000 feet, while far beneath them the gray waters of the Atlantic were rolling and tossing in a gale-force wind. But the westerly wind which had been a problem on the takeoff, was now their friend, boosting them along at more than 100 miles an hour. With both engines running smoothly and the getaway a success, Brown decided to run the wireless apparatus through a few tests, just to make sure it was working properly.

"Say, Jack," he said over the telephone, "I'm going to send St. John's a few words of greeting."

"Tell them we'll be across in sixteen hours," answered Alcock, "if this wind keeps us in its lap."

Brown began tapping out the Vimy's own call letters, DKG, but as he worked the key, the spark grew weaker and weaker until there was no flashing blue light at all. It was a difficult problem, for the wireless generator received its energy from a small, wind-driven propeller located under the forward fuselage and it couldn't be seen from the cockpit.

Brown told his companion about the trouble and, holding tightly to the struts, he climbed out on the wing for a close look at the four-bladed wireless propeller, despite Alcock's emphatic warning against taking such a risk. With the wind screaming around him, Brown crouched on the lower wing, and peering underneath the yellow fuselage, he saw a propeller with only one blade — the other three were gone, probably sheared away by the severe rocking of the plane on the takeoff. For the rest of the flight, their wireless would be out of commission and beyond hope of repair. They would have no way of sending or receiving messages, no way of getting their bearings from ships at sea, and all through the tense hours that lay ahead, they would be completely isolated from the world.

Throughout the early evening hours they flew between layers of clouds above and below, with the ocean completely shut off from view. Then, shortly after 7 P.M., more trouble. The starboard motor began coughing and stuttering so badly that both fliers were thoroughly alarmed as they leaned out to look at the motor. A large chunk of the exhaust pipe had split away, and, as they watched, it changed from red- to white-hot, then gradually crumpled as it grew softer in the intense heat. Three cylinders of the starboard engine were throwing their exhaust fumes straight into the air without benefit of the usual outlet, but throttling down the starboard motor was out of the question, and the exhaust problem was with them to stay.

Brown had brought along a small bubble sextant and was relying entirely on this for directional navigation. It was still evening, and he was taking sights on the stars and moon when suddenly the Vimy plowed into a thick bank of fog. Shouting into the intercom, he told Alcock to go higher, and when they reached 12,000 feet the stars once again were there to guide them.

The hours passed monotonously, and soon it was well past midnight with the moon still shining brightly above them, coloring the clouds with tinges of silver, gold, and red. Below, they could see the Vimy's shadow moving across the layer of clouds that covered the ocean. Twice during the night they polished off a quick sandwich along with a few shots of brandy, and they fought constantly to ignore the drone of the motors which almost lulled them to sleep.

Just as the sun was beginning to rise, the plane ran into another wall of fog, and they had the sensation of flying inside a bottle of milk. With nothing but the white mist around them, they lost their sense of balance and a glance at the instrument panel showed the plane was not on an even keel, and might even be flying upside down. Even worse, it was plunging rapidly through the clouds with the altimeter dropping to 2,000 — 1,000 — then 500 feet. And the tension was growing in the cockpit.

If the cloud layer reached all the way down to the ocean, Alcock would be unable to see the horizon in time to counteract the spin and avoid a crash. Preparing for the worst, Brown loosened his safety belt and they were less than a hundred feet from the water when, in a moment of flashing light, the plane shot out of the cloud vapor into the clear atmosphere.

But the ocean did not appear below them!

The plane was tilted at an angle, and the line of the horizon seemed to be standing vertically to their view. But Alcock quickly regained his visual equilibrium, and the Vimy responded to the controls. At full throttle, he swung the plane back on a level course even though they were flying a mere 50 feet over the ocean. The danger was past, but they were lucky to be out of that one.

For the next three hours, the Vimy moved in and out of a procession of clouds that enveloped the plane time and again, only to give way to patches of blue. But the clouds soon massed into a black wall and a driving rain began lashing the fuselage. Minutes later it turned to snow, and then to a heavy sleet. Trying to rise above the storm, they reached 9,000 feet when Brown saw the glass face of the gasoline overflow gauge clotted with snow and ice. To guard against carburetor trouble, they had to read this gauge at any given moment, and clearing away the ice and snow would be no simple task. It was fixed to one of the center struts and Brown decided that, once more, it was up to him to climb out of the cockpit.

Holding on to a cross-bracing wire to keep the wind from blasting him off the side, he knelt on the wing and managed to reach up and wipe the sleet from the gauge. Time and again, Brown repeated the performance as the storm continued, and he implored Alcock to keep the plane on a level keel while he was out there on the fuselage.

They kept themselves warm by huddling as far down in the open cockpit as possible, while Alcock took the plane higher and higher, hoping to get above the sleet. And it was 6 A.M. when Brown saw the sun glinting through a gap in the clouds. Even though the horizon was shrouded in fog, he was able to get a reading that showed they were nearing the Irish coast and he scribbled Alcock a message:

"Better go lower; the air will be warmer and we may spot a steamer."

Once again, Alcock was feeling his way down through the clouds, knowing that at any moment the wheels might strike the surface of the sea. Then, the plane suddenly emerged from the cloud bank, and both engines responded as Alcock opened full throttle again. But there was no sign of a ship on the cold, gray ocean below as they roared on toward what they hoped would be the emerald coasts of Ireland.

It was exactly 8:15 A.M., and Brown had just finished screwing on the lid of the thermos flask when Alcock grabbed his shoulder, and pointed. There, looming through the mist, were two tiny specks of land—and Brown put both his charts and the thermos away. His navigation duties were over. They had spotted two islands off the Irish coast, and a moment later the mainland came clearly into view.

Still uncertain of their exact location, they crossed the coastline looking for a railway to follow. But when the masts of the Clifden Wireless Station pierced the sky ahead, they knew they were on the beam. It was time for a decision, and Brown asked a crucial question:

"Shall we land at London, or Clifden?"

Alcock was quick to answer, for he feared that Admiral Kerr and his Handley-Page might even then be roaring up somewhere behind them, anxious to beat them to the prize.

"I think we'd better make it Clifden," he said. "All we have to do to win is reach the British Isles, and it'll lengthen our flying time if we go on to London."

As they circled the wireless aerials of Clifden looking for a spot of land, Alcock saw what appeared to be a level stretch of ground. He brought the Vimy in for a perfect landing and it rolled for a hundred feet or so when — without warning — up it went on its nose.

Despite the benign appearance of their landing site, they had come down in an Irish bog. Brown was uninjured, but the soil of Ireland had risen up to give Alcock a black eye and a pair of badly bruised lips. Even so, as he climbed out of the wreckage a grin spread around his swollen mouth, and with a wink toward the Vimy he managed to say:

"She's a pretty fair old boat, eh what?"

The following morning the New York *Times* told the story in its page one headline:

ALCOCK AND BROWN FLY ACROSS THE ATLANTIC;
MAKE 1,980 MILES IN 16 HOURS, 12 MINUTES;
SOMETIMES UPSIDE DOWN IN DENSE, ICY FOG

King George V soon issued the orders making them Sir John Alcock and Sir Arthur Whitten-Brown, and Lord Northcliffe wrote the check for £10,000, shares of which the two fliers generously gave their mechanics.

Brown married his fiancée, Kathleen Kennedy, and he lived to the age of sixty-two. Sir John Alcock, however, was tragically killed in a plane crash in France only six months later.

Now, in the late 1960s, the trail Alcock and Brown blazed so long ago, against such great odds, is flown 1,589 times weekly by huge planes that can speed at 600 miles an hour. But when Orville Wright, on the evening of June 15, 1919, heard about the first nonstop flight across the Atlantic, he couldn't believe his ears.

"What?" he asked with astonishment, "only sixteen hours! Are you sure?"

Suppose Lindbergh, instead of being a modest, unassuming man of natural grace and dignity, had been a lout? That is just what James Thurber supposed in this story he wrote for *The New Yorker* in 1931. Enter Jack ("Pal") Smurch.

JAMES THURBER

The Greatest Man in the World

Looking back on it now, from the vantage point of 1950, one can only marvel that it hadn't happened long before it did. The United States of America had been, ever since Kitty Hawk, blindly constructing the elaborate petard by which, sooner or later, it must be hoist. It was inevitable that some day there would come roaring out of the skies a national hero of insufficient intelligence, background, and character successfully to endure the mounting orgies of glory prepared for aviators who stayed up a long time or flew a great distance. Both Lindbergh and Byrd, fortunately for national decorum and international amity, had been gentlemen; so had our other famous aviators. They wore the laurels gracefully, withstood the awful weather of publicity, married excellent women, usually of fine family, and quietly retired to private life and the enjoyment of their varying fortunes. No untoward incidents, on a worldwide scale, marred the perfection of their conduct on the perilous heights of fame. The exception to the rule was, however, bound to occur and it did, in July 1937, when Jack ("Pal") Smurch, erstwhile mechanic's helper in a small garage in Westfield, Iowa, flew a secondhand, single-motored Bresthaven Dragon-Fly III monoplane all the way around the world, without stopping.

Never before in the history of aviation had such a flight as Smurch's ever been dreamed of. No one had even taken seriously the weird floating auxiliary gas tanks, invention of the mad New Hampshire professor of astronomy, Dr. Charles Lewis Gresham, upon which Smurch placed full reliance. When the garage worker, a slightly built, surly, unprepossessing young man of twenty-two, appeared at Roosevelt Field in early July 1937, slowly chewing a great quid of scrap tobacco, and announced, "Nobody ain't seen no flyin' yet," the newspapers touched briefly and satirically upon his projected 25,000-mile flight. Aeronautical and automotive experts dismissed the idea curtly, implying that it was a hoax, a publicity stunt. The rusty, battered, secondhand plane wouldn't go. The Gresham auxiliary tanks wouldn't work. It was simply a cheap joke.

Smurch, however, after calling on a girl in Brooklyn who worked in the flap-folding department of a large paper-box factory, a girl whom he later described as his "sweet patootie," climbed nonchalantly into his ridiculous plane at dawn of the memorable seventh of July 1937, spit a curve of tobacco juice into the still air, and took off, carrying with him only a gallon of bootleg gin and six pounds of salami.

When the garage boy thundered out over the ocean the papers were forced to record, in all seriousness, that a mad, unknown young man — his name was variously misspelled — had actually set out upon a preposterous attempt to span the world in a rickety, one-engined contraption, trusting to the long-distance refueling device of a crazy schoolmaster. When, nine days later, without having stopped once, the tiny plane appeared above San Francisco Bay, headed for New York, spluttering and choking, to be sure, but still magnificently and miraculously aloft, the headlines, which long since had crowded everything else off the front page — even the shooting of the governor of Illinois by the Vileti gang — swelled to unprecedented size, and the news stories began to run to twenty-five and thirty columns. It was noticeable, however, that the accounts of the epoch-making flight touched rather lightly upon the aviator himself. This was not because facts about the hero as a man were too meagre, but because they were too complete.

Reporters, who had been rushed out to Iowa when Smurch's plane was first sighted over the little French coast town of Serly-le-Mer, to dig up the story of the great man's life, had promptly discovered that the story of his life could not be printed. His mother, a sullen short-order cook in a shack restaurant on the edge of a tourists' camping ground near Westfield, met all inquiries as to her son with an angry "Ah, the hell with him; I hope he drowns." His father appeared to be in jail somewhere for stealing spotlights and lap robes from tourists' automobiles; his young brother, a weak-minded lad, had but recently escaped from the Preston, Iowa, Reformatory and was already wanted in several Western towns for the theft of money-order blanks from post offices. These alarming discoveries were still piling up at the very time that Pal Smurch, the greatest hero of the twentieth century, blear-eyed, dead for sleep, half-starved, was piloting his crazy junk-heap above the region in which the lamentable story of his private life was being unearthed, headed for New York and a greater glory than any man of his time had ever known.

The necessity for printing some account in the papers of the young man's career and personality had led to a remarkable predicament. It was of course impossible to reveal the facts, for a tremendous popular feeling in favor of the young hero had sprung up, like a grass fire, when he was halfway across Europe on his flight around the globe. He was, therefore, described as a modest chap, taciturn, blond, popular with his friends, popular with girls. The only available snapshot of Smurch, taken at the wheel of a phony automobile in a cheap photo studio at an amusement park, was touched up so that the little vulgarian looked quite handsome. His twisted leer was smoothed into a pleasant smile. The truth was, in this way, kept from the youth's ecstatic compatriots; they did not dream that the Smurch family was despised and feared by its neighbors in the obscure Iowa town, nor that the hero himself, because of numerous unsavory exploits, had come to be regarded in Westfield as a nuisance and a menace. He had, the reporters discovered, once knifed the principal of his high school — not mortally, to be sure, but he had knifed him; and on another occasion, surprised in the act of stealing an altarcloth from a church, he had bashed the sacristan over the head with a pot of Easter lilies; for each of these offences he had served a sentence in the reformatory.

Inwardly, the authorities, both in New York and in Washington, prayed that an understanding Providence might, however awful such a thing seemed, bring disaster to the rusty, battered plane and its illustrious pilot, whose unheard-of flight had aroused the civilized world to hosannas of hysterical praise. The authorities were convinced that the character of the renowned aviator was such that the limelight of adulation was bound to reveal him to all the world as a congenital hooligan mentally and morally unequipped to cope with his own prodigious fame. "I trust," said the secretary of state, at one of many secret Cabinet meetings called to consider the national dilemma, "I trust that his mother's prayer will be answered," by which he referred to Mrs. Emma Smurch's wish that her son might be drowned. It was, however, too late for that — Smurch had leaped the Atlantic and then the Pacific as if they were millponds. At three minutes after two o'clock on the afternoon of July 17, 1937, the garage boy brought his idiotic plane into Roosevelt Field for a perfect three-point landing.

It had, of course, been out of the question to arrange a modest little reception for the greatest flier in the history of the world. He was received at Roosevelt Field with such elaborate and pretentious ceremonies as rocked the world. Fortunately, however, the worn and spent hero promptly swooned, had to be removed bodily from his plane, and was spirited from the field without having opened his mouth once. Thus he did not jeopardize the dignity of this first reception, a reception illumined by the presence of the secretaries of war and the navy, Mayor Michael J. Moriarity of New York, the premier of Canada, Governors Fanniman, Groves, McFeely, and Critchfield, and a brilliant array of European diplomats. Smurch did not, in fact, come to in time to take part in the gigantic hullabaloo arranged at City Hall for the next day. He was rushed to a secluded nursing home and confined to bed. It was nine days before he was able to get up, or to be more exact, before he was permitted to get up. Meanwhile the greatest minds in the country, in solemn assembly, had arranged a secret conference of city, state, and government officials, which Smurch was to attend for the purpose of being instructed in the ethics and behavior of heroism.

On the day that the little mechanic was finally allowed to get up and dress and, for the first time in two weeks, took a great chew of tobacco, he was permitted to receive the newspapermen — this by way of testing him out. Smurch did not wait for questions. "Youse guys," he said — and the *Times* man winced — "youse guys can tell the cockeyed world dat I put it over on Lindbergh, see? Yeh — an' made an ass o' them two frogs." The "two frogs" was a reference to a pair of gallant French fliers who, in attempting a flight only halfway round the world, had, two weeks before, unhappily been lost at sea. The *Times* man was bold enough, at this point, to sketch out for Smurch the accepted formula for interviews in cases of this kind; he explained that there should be no arrogant statements belittling the achievements of other heroes, particularly heroes of foreign nations. "Ah, the hell with that," said Smurch. "I did it, see? I did it, an' I'm talkin' about it." And he did talk about it.

None of this extraordinary interview was, of course, printed. On the contrary, the newspapers, already under the disciplined direction of a secret directorate created for the occasion and composed of statesmen and editors, gave out to a panting and restless world that "Jacky," as he had been arbitrarily nicknamed, would consent to say only that he was very happy and that anyone could have done what he did. "My achievement has been, I fear, slightly exaggerated," the *Times* man's article had him protest, with a modest smile. These newspaper stories were kept from the hero, a restriction which did not serve to abate the rising malevolence of his temper. The situation was, indeed, extremely grave, for Pal Smurch was, as he kept insisting, "rarin' to go." He could not much longer be kept from a nation clamorous to lionize him. It was the most desperate crisis the United States of America had faced since the sinking of the *Lusitania.*

On the afternoon of the twenty-seventh of July, Smurch was spirited away to a conference room in which were gathered mayors, governors, government officials, behaviorist psychologists, and editors. He gave them each a limp, moist paw and a brief unlovely grin. "Hah ya?" he said. When Smurch was seated, the mayor of New York arose and, with obvious pessimism, attempted to explain what he must say and how he must act when presented to the world, ending his talk with a high tribute to the hero's courage and integrity. The mayor was followed by Governor Fanniman of New York, who, after a touching declaration of faith, introduced Cameron Spottiswood, Second Secretary of the American Embassy in Paris, the gentleman selected to coach Smurch in the amenities of public ceremonies. Sitting in a chair, with a soiled yellow tie in his hand and his shirt open at the throat, unshaved, smoking a rolled cigarette, Jack Smurch listened with a leer on his lips. "I get ya, I get ya," he cut in, nastily. "Ya want me to ack like a softy, huh? Ya want me to ack like that ———— ———— baby-faced Lindbergh, huh? Well, nuts to that, see?" Everyone took in his breath sharply; it was a sigh and a hiss. "Mr. Lindbergh," began a United States senator, purple with rage, "and Mr. Byrd—" Smurch, who was paring his nails with a jackknife, cut in again. "Byrd!" he exclaimed. "Aw fa God's sake, dat big—" Somebody shut off his blasphemies with a sharp word. A newcomer had entered the room. Everyone stood up, except Smurch, who, still busy with his nails, did not even glance up. "Mr. Smurch," said someone sternly, "the President of the United States!" It had been thought that the presence of the chief executive might have a chastening effect upon the young hero, and the former had been, thanks to the remarkable cooperation of the press, secretly brought to the obscure conference room.

A great, painful silence fell. Smurch looked up, waved a hand at the President. "How ya comin'?" he asked, and began rolling a fresh cigarette. The silence deepened. Someone coughed in a strained way. "Geez, it's hot, ain't it?" said Smurch. He loosened two more shirt buttons, revealing a hairy chest and the tattooed word "Sadie" enclosed in a stenciled heart. The great and important men in the room, faced by the most serious crisis in recent American history, exchanged worried frowns. Nobody seemed to know how to proceed. "Come awn, come awn," said Smurch. "Let's get the hell out of here! When do I start cuttin' in on de parties, huh? And what's they goin' to be *in* it?" He rubbed a thumb and forefinger together meaningly. "Money!" exclaimed a state senator, shocked, pale. "Yeh, money," said Pal, flipping his cigarette out of a window. "An' big money." He began rolling a fresh cigarette. "Big money," he repeated, frowning over the rice paper. He tilted back in his chair, and leered at each gentleman, separately, the leer of an animal that knows its power, the leer of a leopard loose in a bird-and-dog show. "Aw fa God's sake, let's get some place where it's cooler," he said. "I been cooped up plenty for three weeks!"

Smurch stood up and walked over to an open window, where he stood staring down into the street, nine floors below. The faint shouting of newsboys floated up to him. He made out his name. "Hot dog!" he cried, grinning, ecstatic. He leaned out over the sill. "You tell 'em, babies!" he shouted down. "Hot diggity dog!" In the tense little knot of men standing behind him, a quick, mad impulse flared up. An unspoken word of appeal, of command, seemed to ring through the room. Yet it was deadly silent. Charles K. L. Brand, secretary to the mayor of New York City, happened to be standing nearest Smurch; he looked inquiringly at the President of the United States. The President, pale, grim, nodded shortly. Brand, a tall, powerfully built man, once a tackle at Rutgers, stepped forward, seized the greatest man in the world by his left shoulder and the seat of his pants, and pushed him out the window.

"My God, he's fallen out the window!" cried a quick-witted editor.

"Get me out of here!" cried the President. Several men sprang to his side and he was hurriedly escorted out of a door toward a side entrance of the building. The editor of the Associated Press took charge, being used to such things. Crisply he ordered certain men to leave, others to stay; quickly he outlined a story which all the papers were to agree on, sent two men to the street to handle that end of the tragedy, commanded a senator to sob and two congressmen to go to pieces nervously. In a word, he skillfully set the stage for the gigantic task that was to follow, the task of breaking to a grief-stricken world the sad story of the untimely, accidental death of its most illustrious and spectacular figure.

The funeral was, as you know, the most elaborate, the finest, the solemnest, and the saddest ever held in the United States of America. The monument in Arlington Cemetery, with its clean white shaft of marble and the simple device of a tiny plane carved on its base, is a place for pilgrims, in deep reverence, to visit. The nations of the world paid lofty tributes to little Jacky Smurch, America's greatest hero. At a given hour there were two minutes of silence throughout the nation. Even the inhabitants of the small, bewildered town of Westfield, Iowa, observed this touching ceremony; agents of the Department of Justice saw to that. One of them was especially assigned to stand grimly in the doorway of a little shack restaurant on the edge of the tourists' camping ground just outside the town. There, under his stern scrutiny, Mrs. Emma Smurch bowed her head above two hamburger steaks sizzling on her grill—bowed her head and turned away, so that the Secret Service man could not see the twisted, strangely familiar, leer on her lips.

"The R-101 is as safe as a house," declared Lord Thomson shortly before boarding the 777-foot-long airship bound for India, "except for the millionth chance." In fact, because political considerations were permitted to overrule aviation sense, the chance of disaster was far greater than one in a million, as John Toland recounts in this section of his 1957 book, *Ships in the Sky*. When the R-101 crashed and then exploded in northern France, its doom spelled that of the entire British airship program. The R-101's proven predecessor, the R-100, was hacked apart with axes, flattened by a steamroller, and sold for scrap.

JOHN TOLAND

The Millionth Chance

In the spring of 1929, Eckener, always a master showman, announced that in addition to its Atlantic and Mediterranean flights, the *Graf* would that year make a round-the-world tour. Almost immediately all seats for the trip were snapped up. Amid a flood of publicity the zeppelin left Lakehurst at 11:40 P.M., August 29, with an international passenger list of distinction. Twenty-one days, five hours, and fifty-four minutes later, the ship returned after circling the world at an average speed of 70.7 miles an hour. Colonel Charles Lindbergh was among the many thousands who greeted the airship at Lakehurst.

Eckener was lionized, invited to Washington to meet President Hoover, given a tumultuous ticker-tape welcome in New York City. Mayor Jimmy Walker called him "one of the greatest living men in the civilized world." Those who doubted the practicability of the dirigible were silenced.

Within a few months three countries were competing hotly for supremacy in the rigid-dirigible field. At Akron, Ohio, the Goodyear-Zeppelin Company had built the largest hangar in the world to house the two new Navy airships that would both be almost twice as big as the *Graf*. The hangar itself was a thing of wonder. Niagara Falls could have fit inside the colossal room. Strangely shaped like an egg cut in two, each of its great doors looked like a quarter of an orange. The two ships, the ZR-4 and ZR-5, would both have a capacity of 6,500,000 cubic feet and featured engine rooms inside the hull with propellers, on outriggers, that could be pitched from their regular position to drive the ship up or down.

Germany, naturally, was still ahead of the other countries. A vast new passenger ship was being designed with a lifting power of almost half a million pounds and an 8,000-mile cruising radius.

England was also in the race. The two immense commercial airships, ordered built by the Air Ministry for service to the furthest reaches of the Empire, were nearing completion.

This English program had started in 1924. The contract for one dirigible, the R-100, was given to Vickers, and work was started at the deserted hangar in Howden. The second ship, the R-101, was to be built by the same team that designed the R-38 (the ZR-2). Although the R-38 had broken in two over the Humber River with the loss of many lives, and although the subsequent inquiry proved that little effort had been expended to resolve the aerodynamic problems peculiar to airship design, none of the designers was fired — or even censured.

From the first there was bitter rivalry between Lieutenant Colonel Victor Charles Richmond, designer of the R-101, and Mr. B. N. Wallis, designer of the R-100. During the five years of construction neither man visited the other, or even wrote each other about their common problems. The government-built ship, the R-101, which was in the works at Cardington, received most of the publicity. Vickers could hardly compete with the powerful public relations of the Air Ministry.

But the publicity backfired. Once a new device was designed and built, it had to be installed on the R-101 whether it was suitable or not. At Howden the Vickers people discovered that the proposed diesel engines would make their ship too heavy, so they changed to Rolls-Royce gasoline engines. The R-101 had the same weight problem. But because so many articles had been written about the safety features of their diesels, the change was disapproved by the Air Ministry — even though Colonel Richmond strongly urged it.

The R-101 was finished first. Her maiden flight took place on October 14, 1929. The large crowd that came somewhat skeptically stayed to praise. For the government-built craft was beautiful to behold. But in the air she proved to be dangerously underpowered, with a useful lift of only thirty-five tons.

The Vickers ship was tried out the middle of December. The R-100 was 709 feet long and 130 feet in diameter. Since she had fewer longitudinal girders, she lacked the outward beauty of her sister ship. But the R-100 had one virtue — she could fly. Although she, too, was overheavy (her lift was 57 tons instead of the estimated 64 tons), she reached 81 miles an hour — a remarkable speed for six Rolls-Royce Condor motors that had already seen service in the RAF since 1925.

Now the rivalry between the two staffs became even keener. The effects of the American depression were being felt in England, and it was obvious the entire airship program would have to be curtailed. This meant that the designers of the losing ship would lose their jobs.

In desperation the Cardington builders cut the R-101 in two and inserted a new bay to give their ship more lift. On June 2, 1930, the rebuilt dirigible came out of her shed and went on the tower. Her entire cover began rippling from bow to stern, and within minutes a great tear ninety feet long appeared along the top of the ship. Riggers quickly repaired the damage with tapes. But the next day there was another split of forty feet.

The new troubles of the R-101 were withheld from the public, and on June 28 the dirigible was pulled out of her hangar to be shown off at the Hendon Air Pageant. The ship rose easily enough from her home field and reached Hendon without trouble.

But on the return flight to Cardington a few hours later she went into a steep 500-foot dive. Height Coxswain "Mush" Oughton quickly brought the ship's nose up. After a long, slow climb the R-101 was back to her flying height, 1,200 feet. A few minutes later the ship dove again. Once more the coxswain, after a struggle, brought the ship to a safe altitude.

The officer on watch, Captain George Meager, was alarmed. Regularly assigned to the R-100 as first officer, he had never before been in a dirigible that acted like the R-101. It was, he later admitted, the first time he'd "ever had the wind up in an airship."

Sweat was now streaming down the height coxswain's face. "It's as much as I can do to hold her up, sir," he said.

Suddenly the ship dove a third time. Meager grabbed the water-ballast toggles. Over a ton of water spilled out. The R-101 nosed up.

"She's much easier now," said Oughton with relief.

A moment later Flight Lieutenant H. Carmichael Irwin, captain of the ship, came into the control car. He was annoyed by the release of the water ballast. Meager explained that the ship was heavy, but Irwin insisted that air bumps had caused the dives.

Half an hour later the ship landed safely at Cardington. On inspection, more than sixty small holes were found in the hydrogen cells. The R-101, unlike German and American airships, had no network dividing each gas bag. The single-ply cells not only had rubbed against each other but had chafed against girders as well.

The final test for each of the competing English dirigibles was to be a long, spectacular thip. The R-100 was scheduled to fly to Canada and back and the R-101 to India. The R-101 designers, realizing that drastic action had to be taken to make their ship lighter, finally decided to rebuild the new bay. At this time they unofficially suggested to their rivals at Howden that it would be wiser to postpone both overseas trips until the next year.

The R-100 builders at Howden, smarting over the many real and supposed indignities of the past five years, refused. Their ship was ready for the Canada flight, and they intended to make it. Secretly, of course, they were delighted at their rivals' woes. On July 29, 1930, the R-100 left England.

Although one of the secondhand engines went out of commission, the trip over the ocean was completed without mishap. But while the ship was cruising along the St.

THE **R.101** WAS "AS SAFE AS A HOUSE..."

...INSISTED AIR MINISTER LORD THOMPSON AS HE BOARDED ENGLAND'S NEWEST AIRSHIP ON OCTOBER 4, 1930, BUT THE DAY BEFORE HE'D PRUDENTLY MADE HIS WILL.

THAT NIGHT, NEAR BEAUVAIS, FRANCE, THE SINGULARLY NON-AIRWORTHY R-101 BEGAN TO LOSE HEIGHT, SUDDENLY DIVED, HIT A LOW HILL AND BURST INTO FLAMES, KILLING THOMPSON AND 47 OTHERS.

Lawrence River, a large section of fabric ripped off the fins. The accident was similar to the *Graf Zeppelin*'s near-disaster in mid-ocean; and, as with the *Graf*, repairs were made in mid-air. Then, just after the R-100 had passed over Quebec, a storm appeared dead ahead. Although Squadron Leader Booth was the ship's captain, Major G. H. Scott, an extremely able but overly daring man, ordered the helmsman to steer straight into the storm rather than skirt it in the Eckener manner. In the heavy weather more fabric ripped off. Even so the ship arrived safely at Montreal and was given a typical New World welcome. Posters of the R-100 were plastered all over the city, and a song was written in Captain Booth's honor.

After repairs were made, the dirigible went on a twenty-four-hour tour of Canada, impressing the colonials with the technology of the mother country. Twelve days after its arrival in Montreal, the R-100 started back across the Atlantic on five engines, touching down at Cardington after a speedy trip of fifty-seven and a half hours. Only a few hundred people were on hand to greet the triumphant fliers.

In spite of the skimpy reception, and not withstanding the ship's two casualties, it was quite evident that the R-100's round-trip crossing of the Atlantic had been a substantial success. It was a success that forced the designers of the R-101 to make a difficult decision: to admit that their rivals had the better ship — and lose their jobs — or to fly to India. The choice was India.

At this point a complication arose to make the choice binding. Lord Thomson, the energetic Secretary of State for Air, announced that he would fly on the R-101 so that he could keep several political engagements in India. These meetings were scheduled for September. "I must insist on the programme for the Indian flight being adhered to," he wrote the Cardington officials, "as I have made my plans accordingly."

The desperate R-101 builders had to tell their chief that this would be impossible — the new bay would not be repaired until October 1. Impatiently Thomson laid out a new schedule. They would leave England on October 4, reach the Karachi mast on the 9th. They would leave Karachi on the 13th and arrive back in England on October 18. Such a rigid schedule for an experimental ship was more than impractical — it was foolhardy. But Thomson, who was slated to be the next Viceroy of India, held firm. He wanted to fly to his new empire in the most dramatic style.

Instead of protesting, the officials at Cardington rushed the repairs.

Late in September, a few days before the scheduled trip to India, one of the engineer-builders of the R-100 visited Cardington to see his good friend, Squadron Leader Booth. The engineer was Nevil Shute Norway — later and better known as the novelist Nevil Shute.

The captain of the R-100 showed Norway a piece of outer-cover fabric. "What do you think of that?" he asked.

The fabric, which looked like scorched brown paper, crumpled into flakes when Norway squeezed it. He was horrified.

"That's off R-101," Booth told him. Rubber solution, applied to strengthen the fabric, had produced this strange and ominous reaction.

"I hope they've got all this stuff off the ship," said Norway.

"They *say* they have," replied Booth.

On October 1 the R-101, now 777 feet long, was brought out of the hangar for a trial. Soon after the ship took to the air the oil cooler of one engine failed. Although a flight of sixteen hours was logged, it was impossible to give the engines a full power trial.

Immediately the abbreviated test flight was over, Thomson ordered the ship started the next evening. The designers said the crew needed a rest. They compromised on the evening of October 4.

Then it was discovered that the ship lacked a certificate of aeronautics, a requirement for any flight over foreign land. Since there was no time for a thoroughgoing inspection, the ship received a quick once-over. The certificate was written out in the Air Ministry and handed to Captain Irwin as passengers and crew filed aboard.

On that Saturday evening the great majority of a large crowd stared in awe at the majestic dirigible as she parted from the mast. As for the few who were privy to the expediencies of the past five years, their awe was tempered with foreboding. In silence they watched the great ship, its fat sides scarred with the incisions of its recent operation, disappear slowly into the darkness.

The innocent majority cheered. Hadn't Lord Thomson told the world that "the R-101 is as safe as a house, except for the millionth chance"?

The biggest airship in the world left Cardington at 6:36 P.M. with six passengers, including Lord Thomson; his valet, James Buck; the Director of Airship Development, Wing Commander Colmore; six officials of the Royal Airship Works, headed by the ship's designer, Colonel Richmond; and a crew of forty-two.

Although nine and a half tons of water had been taken on, before the ship left the mast four tons were dumped to compensate for the heavy passenger and fuel load. All day the barometer had been falling threateningly.

Over Hitchin rain clouds loomed ahead, and strong, gusty wings made the R-101 roll and pitch heavily.

"I never knew her to roll so much," Squadron Leader Rope told Henry Leech, foreman engineer of the Royal Airship Works. "She's moving more like a seagoing ship than an aircraft."

A moment later the aft engine stopped. "It's the main oil pressure, sir," Engineer Arthur Bell told his chief engineer, W. R. Gent. Both men struggled unsuccessfully with the balky motor. Then civilian expert Leech was called. The three worked in the cramped gondola; they were working there at 8:08 P.M., as the ship sailed sluggishly over London. Rain began beating on top of the dirigible and dripped into the water-recovery system.

A revised weather forecast now came over the radio: there would be winds of forty to fifty miles an hour over northern France. Major Scott, who was in charge of flying operations, glanced at the storm warning. But he didn't order the ship to return to

Cardington. He headed the ship toward Paris — toward the heart of the brewing storm. He took the millionth chance.

The R-101, still flying with one motor out, crossed the English coast near Hastings at 9:35 P.M. No one in the control car was worried. A wire was sent to Cardington: "Ship is behaving well generally and we have already begun to recover water ballast."

In mid-channel the aft engine was finally repaired. As Leech wearily climbed the ladder into the belly of the ship, he looked down and saw whitecaps. He estimated, somewhat apprehensively, that they were only 700 feet above the water.

The first officer, Lieutenant Commander N. G. Atherstone, was worried. After looking at the altimeter he took the elevator wheel from Height Coxswain Oughton. In a minute the ship soared up to 1,000 feet.

Atherstone handed the wheel back to the coxswain. "Don't let her go below one thousand feet," he warned. Perhaps the commander was thinking of an entry he'd made in his diary just the night before. "I feel that that thing called 'Luck,'" he'd written, "will figure conspicuously in our flight. Let's hope for good luck and do our best."

At eleven o'clock the watch was changed, and twenty-six minutes later the French coast was crossed at Pointe de St.-Quentin. The winds had increased to 35 miles an hour, and the dirigible was having trouble making headway.

But despite the buffeting Major Scott was well pleased with the R-101's performance. "After an excellent supper," the ship wired Cardington, "our distinguished passengers smoked a final cigar and, having sighted the French coast, have now gone to rest after the excitement of their leave-taking."

All was serene aboard the R-101, and the crew settled down for their long, tedious watch-keeping routine. By 1:00 A.M. only Chief Engineer Gent and civilian engineer Henry Leech were in the huge, luxurious smoking room. It was the first smoking room on any aircraft. It had been built in spite of the threat of hydrogen.

Captain Irwin, the ship's nominal skipper, dropped in for a cigarette a moment later. He told the two engineers that the aft engine had settled down and was running smoothly. Then he crushed out his butt and went forward to the control car. Gent yawned, said good night, and headed for the officers' quarters.

Leech, nervous and proud as a mother hen, now rose and began making the rounds of his engines. All five diesels were working as well as they had in extensive bench tests. But he still was a bit worried: no full-power test had been made since the insertion of the new bay. Unable to sleep, Leech went back to the smoking room. He mixed himself a drink, lit a cigarette, and then sat on a long, comfortable settee. He tried to relax but he couldn't.

The ship was just then passing over Poix Aerodrome, halfway between Abbéville and Beauvais. Louis Maillet, the resident in charge of the field, heard the dirigible's motors and looked out the window of his house. Although it was too cloudy to see the outline of the R-101, Maillet could distinguish a line of white lights moving slowly against the strong wind. The ship seemed only 100 yards above the field, which was on a plateau 200 yards high. Maillet wondered why the R-101 was flying less than half her own length from the ground.

In the control car the men noticed how close they were to the ground; yet the altimeter, which was set at sea level, recorded 1,000 feet. The navigator was confused. He wrote out a message for Le Bourget airfield, 40 miles away. "What is my true bearing?" he asked.

At 1:51 A.M. Le Bourget finally worked out the ship's position: five-eighths of a mile north of the field at Beauvais. A minute later the British dirigible acknowledged receipt of the information.

It was the last message from the R-101.

By this time the charge-hand, G. W. Short, was waking up the men for the two-o'clock watch. He shook engineer Alfred Cook. Yawning, Cook looked up at the gas bag above him — 8 A, the new one. It was surging about him; he'd never seen it so pendulous before. But he thought little of it. He put on his shoes and started down the keel toward the port amidship engine. He knew his partner, Blake, must be getting tired.

Another engineer, Victor Savory, was out of his bunk instantly. He stopped up his ears with cotton wool and plasticene — the noise of the engine drove him crazy. In two minutes he had climbed down into the starboard amidship engine room.

His partner, Hastings, gave Savory the thumbs-up signal to show everything was fine. "Get your carcass out quick!" called Savory cheerfully. "Get some sleep."

Short had a harder time rousing John Henry Binks, one of the aft engineers.

"How's my bloody engine behaving?" asked Binks wearily.

Short told him it was doing fine. "Now get a move on," he added impatiently.

Binks, who hadn't slept well because of the pitching and rolling, sauntered to the crew space and poured himself a cup of hot cocoa. Old Bell wouldn't mind a wait of a few extra minutes.

The first one in Beauvais to be awakened by the motors of the R-101 was Jules Patron, secretary of the Beauvais Police Station. He jumped out of bed, put on a pair of trousers and a jacket, and ran out into the rainy marketplace. He knew it was the great British airship. The papers had been full of news about its exciting trip to India by way of Ismailia. Even in the dark and storm he could see the outline of the dirigible. It was heading south. There were three green lights on the side. To get a better view Patron hurried home and climbed to the third story.

Julien Lechat, the jeweler, was waiting for the ship, too. He had worked in his shop until 1:00 A.M., for the next day was Saturday, market day. He was taking off his clothes when he heard the low drum of motors. He looked out his bedroom window and saw the ship "fighting the elements" just above the tower of St. Etienne. The dirigible was rolling and dipping. The wind blew in great squalls and the rain beat against Lechat's windowpanes like little bullets.

A few blocks away, Madame Julie Sostier, concierge at the Palais de Justice, awakened quickly. She was a mass moving sideways slowly and with great difficulty. From her window she could see two white lights in front of the dirigible, a green light on the left, and a red one on the right. She was the only one in town who thought the ship was in trouble. It was so low she felt certain it was going to crash right in the center of her beloved town.

Lights began flashing on all over Beauvais. Everyone in the Woillez house was awake, and the children were wildly excited. The clock of St. Etienne struck two as the Bards, who lived across from the church, woke up. Madame Bard, knowing it was the R-101, threw open the window and peered out. She saw the ship just behind the tennis courts. It seemed to be about 200 yards above the ground.

In the port amidship gondola, Cook had already relieved Blake. Everything was all right, Blake had said, but Cook was the kind who liked to see for himself. He picked up his torch and flashed a beam of light along each side of his engine. Then he checked the instruments. Finally he looked at the log. The engine had been at cruising speed since seven o'clock. Everything *was* all right.

Binks was still drinking hot cocoa in the crew space. Then Short caught sight of him and told him to get cracking. The engineer started down the keel. After about 150 feet he reached the ladder leading down into the aft gondola. It was pitch black, and the violent wind tore at him as he climbed into the engine room.

"You're late!" Arthur Bell said good-naturedly.

"Only three minutes," replied Binks.

"Five!" Bell said, and they gossiped for a minute about the temperature and pressures. Suddenly the ship went into a dive. Binks, who was leaning against the oil cooler, sat down with a thump. Bell, facing aft, fell against the starting engine.

"She's really got a nasty angle on!" thought Bell. But he didn't say anything—he just looked at Binks on the floor.

To Savory, in the port amidship engine room, the dip was "nothing to write home about." It wasn't even worth putting down in the log. In the starboard gondola, Alf Cook also thought it was just a "slight diving attitude."

But Electrician Arthur Disley, in his bunk just above the two amidship engines, was wakened out of a sound sleep by the dive. And in the smoking room forward, civilian engineer Leech's settee slid across the floor and banged against the forward bulkhead. A Sparklet Syphon and several glasses toppled off the table. As the ship leveled off somewhat, Leech staggered to his feet. He leaned over to pick up the debris.

Rigger Church, who a few minutes before had been relieved from his watch far forward, was walking back to the crew space when the ship dove. He grabbed a hand line for support.

"Release emergency ballast!" an officer shouted up from the control car to the passing rigger.

Church knew they were in trouble. He turned and ran forward along the keel to jettison the half-ton of water in the nose.

Leech by now had pulled the table from the wall and replaced the syphon and glasses. As he sat down on the settee again he could hear the telegraph bells in the control car ringing. Something was up.

The men in the control car already knew that the ship was doomed. Even though Height Coxswain Oughton had the wheel in a hard "up" position, the elevators weren't responding. Another dive was about to come. First Officer Atherstone ordered Chief Coxswain Hunt to go into the keel and rouse the sleeping crew to stand by for a crash.

The chief coxswain ran up into the ship and back toward the enlisted men's sleeping quarters. Arthur Disley was still in his bunk when Hunt shouted, "We are down, lads!"

Before Disley could ask him what was wrong, Hunt was running toward the tail of the dirigible, repeating his warning. Now Disley was aware of the jangle of telegraph bells in the control car.

Binks heard the aft engine telegraph ringing just as he got off the floor. The light was flashing a "slow" signal. His partner, Bell, instantly rang back an answer and then threw the motor into "slow."

Cook in the port engine room, had received the unusual "slow" signal during the first dive, and his motor was already slowed down. He knew something serious had happened. As he peered out the gondola window, the ship went into another dive, steeper than the first. Hanging onto the door, Cook looked down but he could see only blackness below. He wondered how high up they were and what was wrong.

The second dive threw both Bell and Savory against their starting engines. Savory still didn't think much had happened. He looked up at his telegraph for an order. There was no blinking light. He figured that it was just the usual rough ride that followed a watch change. During sticky weather it usually took a few minutes for the new height coxswain to get the feel of the airship.

Few of those watching the dirigible noticed the first dive, but they all saw the second. From his window Louis Pettit, a wine merchant of Allonne, had been following the ship's shaky route for several minutes. As the R-101 passed between the villages of Allonne and Bongenoult he noticed that it "put out its fires." A moment later the lights came on again, and the dirigible started dipping toward the ground. Fernand Radel, a neighboring farmer, also saw the lights go off and on. Neither man realized the ship was only yawing in the heavy wind.

Louis Tellier, a shepherd of Bongenoult, was lying on his cot watching the dirigible fly over his little hut. The row of lights reminded him of a passing railroad train. Then the ship started to dive. Instinctively he turned his head to see if his sheep were safe.

Alfred Rabouille, who worked in a button factory by day and caught rabbits by night, was setting snares near the edge of a small woods when he heard the roar of engines. He looked up through the fine rain and saw a huge airship wobbling unsteadily. Suddenly it went into a dive. As he watched it headed straight for him. He stood rooted to the ground, not knowing what to do.

Electrician Disley, still drugged with sleep, swung his legs out of bed. Hunt's warning had told him the ship was going to crash. He knew he had to turn off the ship's electricity before it touched off an explosion of 5,500,000 cubic feet of hydrogen. With his left hand he punched a button that released one of the two field switches on the switchboard located at the head of his bed. But before he could push the second button he heard a "crushing" noise.

It was the underpart of the bow striking the earth. The dirigible hadn't hit with much impact and it pancaked gently along the ground for sixty feet. Rabouille, only 250 yards away, thought the fliers were in for no more than a little shaking up.

Then came a terrific, blinding explosion. Two smaller explosions followed. The rabbit catcher stared in horror as a huge flare lit up the sky. Then three rapid explosions knocked Rabouille off his feet.

In Beauvais those who were watching from the rooftops had seen the ship dive out of sight behind trees. Then they saw a great flash.

"What is that light?" asked Madame Bard, who was now sitting up in bed. Her husband looked sleepily out the window. A light like daybreak lit up the sky. Seven seconds later came a dull rumble that grew in intensity, shaking the house. Jules Patron, of the police station, could see bits of "flying paper" shoot up in the air in the distance.

The port amidship gondola didn't touch the ground at the first impact. Instinctively Cook turned off his ignition. Then he felt, rather than heard, an explosion. His car seemed to drop, with only a slight bump, onto the ground. He jumped out and ran.

On the starboard side, Engineer Savory had no idea what had happened. Before he could turn off his motor, a vivid flash shot through the gondola's open door, scorching his face and momentarily blinding him.

The aft gondola hit the earth lightly at first. It bounced along until the bottom caved in. Water suddenly poured onto Binks and Bell. They thought the ship had fallen into a river until they saw water falling from a broken tank overhead. Waves of flame were now rushing aft with a great rumble. The two men crouched in the cabin a moment. Then they saw a path cut in the flames by the water and leaped out of the gondola. Covering their faces with wet handkerchiefs, they flopped on the dewy grass and rolled over and over — away from the licking flames.

Civilian Leech was sitting on the smoking room settee when the ship struck the ground. There was little shock. He merely slid down the settee. The lights went out. Then there was an intense flare in the doorway. The ceiling collapsed onto the settee. Leech dropped to his knees and crawled aft. With his hands he tore a path through a thin partition. Then he jumped through the flaming envelope. He was the only civilian who had not been trapped in bed.

Savory did not know how he had escaped from his burning engine room. He could still hear the crackling of flames as he was led to safety by a French peasant.

The bow of the ship was now resting in the woods, the aft section in a meadow. Electrician Disley, dazed and shielding his eyes from the blinding heat of the flames, looked at his ship. The envelope was still burning on some parts of the R-101, but there seemed to be no cover at all on top of the dirigible. Apparently great hunks of the top had ripped off in the two dives.

Cook was fascinated by the blaze. He walked a few yards through the woods so he could look back at the tail. In the searing light he could clearly see the huge elevators. They were jammed in a severe "up" position. Height Coxswain Oughton had obviously held tightly onto the wheel until the last minute in a vain effort to keep the ship's nose up.

In a few minutes townspeople of Allonne and Bongenoult ran onto the field. They did what they could, herding the eight who still lived to safety. Two of these, riggers Church and Radcliffe, soon died from their injuries. Only six survived: Leech, Disley, Cook, Savory, Binks, and Bell.

The crash of the R-101 stirred the British people more deeply than any disaster since the war. In one stroke Secretary of State for Air Thomson, the director of civil aviation, and the elite of Great Britain's airship service had been wiped out.

A great public funeral for the R-101 dead was held in London. The cortege of forty-eight coffins, stretching for many blocks, was a grim spectacle. It marked the end of rigid dirigibles in England.

When Santos-Dumont learned of the tragedy, he declared that the deaths were his responsibility — for he had invented the airship. His sense of guilt, festering since the World War, drove him to hang himself with a necktie. His nephew cut him down just in time.

Visionary, novelist, and historian, Herbert George Wells was one of the great authors of his day, although his star has dimmed somewhat in the years since his death in 1946. His early, pre-1900 novels were pseudoscientific thrillers in the mold of Jules Verne, such as *The Time Machine* and *The War of the Worlds*. These are still read, but the realistic works by which he preferred to be remembered gather dust on the shelves. The short story below is neither scientific nor realistic but simply hilarious.

H. G. WELLS

My First Airplane

My first airplane! What vivid memories of youth that recalls!

Far back it was, in the spring of 1912, that I acquired *Alauda Magna*, the great Lark, for so I christened her; and I was then a slender young man of four-and-twenty, with hair — beautiful blond hair — all over my adventurous young head. I was a dashing young fellow enough, in spite of the slight visual defect that obliged me to wear spectacles on my prominent, aquiline, but by no means shapeless nose — the typical flyer's nose. I was a good runner and swimmer, a vegetarian as ever, an all-wooler, and an ardent advocate of the extremest views in every direction about everything. Precious little in the way of a movement got started that I wasn't in. I owned two motor bicycles, and an enlarged photograph of me at that remote date, in leather skullcap, goggles, and gauntlets, still adorns my study fireplace. I was also a great flyer of war kites, and a voluntary scoutmaster of high repute. From the first beginnings of the boom in flying, therefore, I was naturally eager for the fray.

I chafed against the tears of my widowed mother for a time, and at last told her I could endure it no longer.

"If I'm not the first to fly in Mintonchester," I said, "I leave Mintonchester. I'm your own son, mummy, and that's *me!*"

And it didn't take me a week to place my order when she agreed.

I found one of the old price lists the other day in a drawer, full of queer woodcuts of still queerer contrivances. What a time that was! An incredulous world had at last consented to believe that it could fly, and in addition to the motor-car people and the bicycle people, and so on, a hundred new, unheard-of firms were turning out airplanes of every size and pattern to meet the demand. Amazing prices they got for them too — 350 was cheap for the things! I find 450, 500, 500 *guineas* in this list of mine; and many as capable of flight as oak trees! They were sold, too, without any sort of guarantee, and with the merest apology for instruction. Some of the early airplane companies paid nearly 200 percent on their ordinary shares in those early years.

How well I remember the dreams I had — and the doubts!

The dreams were all of wonder in the air. I saw myself rising gracefully from my mother's paddock, clearing the hedge at the end, circling up to get over the vicar's pear trees, and away between the church steeple and the rise of Withycombe, toward the marketplace. Lord! how they would stare to see me! "Young Mr. Betts again!" they would say. "We *knew* he'd do it."

I would circle and perhaps wave a handkerchief, and then I meant to go over Lupton's gardens to the grounds of Sir Digby Foster. There a certain fair denizen might glance from the window. . . .

Ah, youth! Youth!

My doubts were all of the make I should adopt, the character of the engines I should choose. . . .

I remember my wild rush on my motorbike to London to see the things and give my order, the day of muddy-traffic dodging as I went from one shop to another, my growing exasperation at hearing everywhere the same refrain, "Sold out! Can't undertake to deliver before the beginning of April."

Not me!

I got *Alauda Magna* at last at a little place in Blackfriars Road. She was an order thrown on the firm's hands at the eleventh hour by the death of the purchaser through another maker, and I ran my modest bank account into an overdraft to get her — to this day I won't confess the price I paid for her. Poor little Mumsey! Within a week she was in my mother's paddock, being put together after transport by a couple of not-too-intelligent mechanics.

The joy of it! And a sort of adventurous tremulousness. I'd had no lessons — all the qualified teachers were booked up at stupendous fees for months ahead; but it wasn't in my quality to stick at a thing like that! I couldn't have endured three days' delay. I assured my mother I had had lessons, for her peace of mind — it is a poor son who will not tell a lie to keep his parent happy.

I remember the exultant turmoil of walking round the thing as it grew into a credible shape, with the consciousness of half Mintonchester peering at me through the hedge, and only deterred by our new trespass-board and the disagreeable expression of Snape, our trusted gardener, who was partly mowing the grass and partly on sentry-go with his scythe, from swarming into the meadow. I lit a cigarette and watched the workmen sagely, and we engaged an elderly unemployed named Snorticombe to keep watch all night to save the thing from meddlers. In those days, you must understand an airplane was a sign and a wonder.

Alauda Magna was a darling for her time, though nowadays I suppose she would be received with derisive laughter by every schoolboy in the land. She was a monoplane, and, roughly speaking, a Blériot, and she had the dearest, neatest seven-cylinder 40-horsepower GKC engine, with its GBS flywheel, that you can possibly imagine. I spent an hour or so tuning her up—she had a deafening purr, rather like a machine gun in action—until the vicar sent round to say that he was writing a sermon upon "Peace" and was unable to concentrate his mind on that topic until I desisted. I took his objection in good part, and, after a culminating volley and one last lingering look, started for a stroll round the town.

In spite of every endeavor to be modest I could not but feel myself the cynosure of every eye. I had rather carelessly forgotten to change my leggings and breeches I had bought for the occasion, and I was also wearing my leather skullcap with earflaps carelessly adjusted, so that I could hear what people were saying. I should think I had half the population under fifteen at my heels before I was halfway down the High Street.

"You going to fly, Mr. Betts?" says one cheeky youngster.

"Like a bird!" I said.

"Don't you fly till we comes out of school," says another.

It was a sort of Royal progress that evening for me. I visited old Lupton, the horticulturist, and he could hardly conceal what a great honor he thought it. He took me over his new greenhouse—he had now got, he said, three acres of surface under glass—and showed me all sorts of clever dodges he was adopting in the way of intensive culture, and afterwards we went down to the end of his old flower garden and looked at his bees. When I came out my retinue of kids was still waiting for me, reinforced. Then I went round by Paramors and dropped into the Bull and Horses, just as if there wasn't anything particular up, for a lemon squash. Everybody was talking about my airplane. They just shut up for a moment when I came in, and then burst out with questions. It's odd nowadays to remember all that excitement. I answered what they had to ask me and refrained from putting on any side, and afterward Miss Flyteman and I went into the commercial room and turned over the pages of various illustrated journals and compared the pictures with my machine in a quiet, unassuming sort of way. Everybody encouraged me to go up—everybody.

I lay stress on that because, as I was soon to discover, the tides and ebbs of popular favor are among the most inexplicable and inconsistent things in the world.

I particularly remember old Cheeseman, the pork butcher, whose pigs I killed, saying over and over again, in a tone of perfect satisfaction, "You won't 'ave any difficulty in going *up*, you won't. There won't be any difficulty 'bout going *up*." And winking and nodding to the other eminent tradesmen there assembled.

I *hadn't* much difficulty in going up. *Alauda Magna* was a cheerful lifter, and the roar and spin of her engine had hardly begun behind me before she was off her wheels — snap, snap, they came up above the *ski* gliders — and swaying swiftly across the meadows toward the vicarage hedge. She had a sort of onward roll to her, rather like the movement of a corpulent but very buoyant woman.

I had just a glimpse of brave little mother, trying not to cry, and full of pride in me, on the veranda, with both the maids and old Snape beside her, and then I had to give all my attention to the steering wheel if I didn't want to barge into the vicar's pear trees.

I'd felt the faintest of tugs just as I came up, and fancied I heard a resounding whack on our new Trespassers will be Prosecuted board, and I saw the crowd of people in the lane running this way and that from my loud humming approach; but it was only after the flight was all over that I realized what that fool Snorticombe had been up to. It would seem he had thought the monster needed tethering — I won't attempt to explain the mysteries of his mind — and he had tied about a dozen yards of rope to the end of either wing and fixed them firmly to a couple of iron guy posts that belonged properly to the Badminton net. Up they came at the tug of *Alauda*, and now they were trailing and dancing and leaping along behind me, and taking the most vicious dives and lunges at everything that came within range of them. Poor old Templecom got it hottest in the lane, I'm told — a frightful whack on his bald head; and then we ripped up the vicar's cucumber frames, killed and scattered his parrot, smashed the upper pane of his study window, and just missed the housemaid as she stuck her head out of the upper bedroom window. I didn't, of course, know anything of this at the time — it was on a lower plane altogether from my proceedings. I was steering past his vicarage — a narrow miss — and trying to come round to clear the pear trees at the end of the garden — which I did with a graze — and the trailers behind me sent leaves and branches flying this way and that. I had reason to thank Heaven for my sturdy little GKCs.

Then I was fairly up for a time.

I found it much more confusing than I had expected; the engine made such an infernal whir-r-row for one thing, and the steering tugged and struggled like a thing alive. But I got her heading over the marketplace all right. We buzzed over Stunt's the greengrocer, and my trailers hopped up his back premises and made a sanguinary mess of the tiles on his roof, and sent an avalanche of broken chimney pot into the crowded street below. Then the thing dipped — I suppose one of the guy posts tried to anchor for a second in Stunt's rafters — and I had the hardest job to clear the Bull and Horses stables. I didn't, as a matter of fact, completely clear them. The ski-like alighting runners touched the ridge for a moment and the left wing bent against the top of the chimney stack and floundered over it in an awkward, destructive manner.

I'm told that my trailers whirled about the crowded marketplace in the most diabolical fashion as I dipped and recovered, but I'm inclined to think all this part of the story has been greatly exaggerated. Nobody was killed, and I couldn't have been half a minute from the time I appeared over Stunt's to the time when I slid off the stable roof and in among Lupton's glass. If people had taken reasonable care of themselves instead of gaping at me, they wouldn't have got hurt. I had enough to do without pointing out to people that they were likely to be hit by an iron guy post which had seen fit to follow me. If anyone ought to have warned them it was that fool Snorticombe. Indeed, what with the incalculable damage done to the left wing and one of the cylinders getting out of rhythm and making an ominous catch in the whirr, I was busy enough for anything on my own private personal account.

I suppose I am in a manner of speaking responsible for knocking old Dudney off the station bus, but I don't see that I can be held answerable for the subsequent evolutions of the bus, which ended after a charge among the market stalls in Cheeseman's shop window, nor do I see that I am to blame because an idle and ill-disciplined crowd chose to stampede across a stock of carelessly distributed earthenware and overturned a butter stall. I was a mere excuse for all this misbehavior.

I didn't exactly fall into Lupton's glass, and I didn't exactly drive over it. I think ricocheting describes my passage across his premises as well as any single word can.

It was the queerest sensation, being carried along by this big, buoyant thing, which had, as it were, bolted with me, and feeling myself alternately lifted up and then dropped with a scrunch upon a fresh greenhouse roof, in spite of all my efforts to get control. And the infinite relief when at last, at the fifth or sixth pounce, I rose — and kept on rising!

I seemed to forget everything disagreeable instantly. The doubt whether after all *Alauda Magna* was good for flying vanished. She was evidently very good. We whirred over the wall at the end, with my trailers still bumping behind, and beyond one of them hitting a cow, which died next day, I don't think I did the slightest damage to anything or anybody all across the breadth of Cheeseman's meadow. Then I began to rise, steadily but surely, and, getting the thing well in hand, came swooping round over his piggeries to give Mintonchester a second taste of my quality.

I meant to go up in a spiral until I was clear of all the trees and things and circle about the church spire. Hitherto I had been so concentrated on the plunges and tugs of the monster I was driving, and so deafened by the uproar of my engine, that I had noticed little of the things that were going on below; but now I could make out a little lot of people, headed by Lupton with a garden fork, rushing obliquely across the corner of Cheeseman's meadow. It puzzled me for a second to imagine what they could think they were after.

Up I went, whirring and swaying, and presently got a glimpse down the High Street of the awful tangle everything had got into in the marketplace. I didn't at the time connect that extraordinary smashup with my transit.

It was the jar of my whack against the weathercock that really stopped my engines. I've never been able to make out quite how it was I hit the unfortunate vane; perhaps the twist I had given my left wing on Stunt's roof spoilt my steering; but, anyhow, I hit the gaudy thing and bent it, and for a lengthy couple of seconds I wasn't by any means sure whether I wasn't going to dive straight down into the marketplace. I got her right by a supreme effort — I think the people I didn't smash might have squeezed out one drop of gratitude for that — drove pitching at the treetops of Withycombe, got round, and realized the engines were stopping. There wasn't any time to survey the country and arrange for a suitable landing place; there wasn't any chance of clearing the course. It wasn't my fault if a quarter of the population of Mintonchester was swarming out over Cheeseman's meadows. It was the only chance I had to land without a smash, and I took it. Down I came, a steep glide, doing the best I could for myself.

Perhaps I did bowl a few people over, but progress is progress.

And I had to kill his pigs. It was a case of either dropping among the pigs and breaking my rush, or going fill tilt into the corrugated iron piggeries beyond. I might have been cut to ribbons. And pigs are born to die.

I stopped, and stood up stiffly upon the framework and looked behind me. It didn't take me a moment to realize that Mintonchester meant to take my poor efforts to give it an Aviation Day all to itself in a spirit of ferocious ingratitude.

The air was full of the squealing of the two pigs I had pinned under my machine and the bawling of the nearer spectators. Lupton occupied the middle distance with a garden fork, with the evident intention of jabbing it into my stomach. I am always pretty cool and quick-witted in an emergency. I dropped off poor *Alauda Magna* like a shot, dodged through the piggery, went up by Frobisher's orchard, nipped over the yard wall of Hink's cottages, and was into the police station by the back way before anyone could get within fifty feet of me.

"Halloa!" said Inspector Nenton, "smashed the thing?"

"No," I said, "but people seem to have got something the matter with them. I want to be locked in a cell."

For a fortnight, do you know, I wasn't allowed to come near my own machine. I went home from the police station as soon as the first excitement had blown over a little, going round by Love Lane and the Chart, so as not to arouse any febrile symptoms. I found mother frightfully indignant, you can be sure, at the way I had been treated. And there, as I say, was I, standing a sort of siege in the upstairs rooms, and sturdy little *Alauda Magna*, away in Cheeseman's fields, being walked round and stared at by everybody in the world but me. Cheeseman's theory was that he had seized her. There came a gale one night, and the dear thing was blown clean over the hedge among Lupton's greenhouses again, and then Lupton sent round a silly note to say that if we didn't remove her she would be sold to defray expenses, going off into a long tirade about damages and his solicitor. So mother posted off to Clamps', the furniture removers at Upnorton Corner, and they got hold of a timber wagon, and popular feeling had allayed sufficiently before that arrived for me to go in person to superintend the removal. There she lay like a great moth above the debris of some cultural projects of Lupton's, scarcely damaged herself except for a hole or so and some bent rods and stays in the left wing and a smashed skid. But she was bespattered with pigs' blood and pretty dirty.

I went at once by instinct for the engines, and had them in perfect going order before the timber wagon arrived.

A sort of popularity returned to me with that procession home. With the help of a swarm of men we got *Alauda Magna* poised on the wagon, and then I took my seat to see she balanced properly, and a miscellaneous team of seven horses started to tow her home. It was nearly one o'clock when we got to that, and all the children turned out to shout and jeer. We couldn't go by Pook's Lane and the vicarage, because the walls are too high and narrow, and so we headed across Cheeseman's meadows for Stokes' Waste and the Common, to get round by that detour.

I was silly, of course, to do what I did—I see that now—but sitting up there on my triumphal car with all the multitude about me excited me. I got a kind of glory on. I really only meant to let the propellers spin as a sort of hurrahing, but I was carried away. Whuz-z-z-z! It was like something blowing up, and behold! I was sailing and plunging away from my wain across the common for a second flight.

"Lord!" I said.

I fully meant to run up the air a little way, come about, and take her home to our paddock, but those early airplanes were very uncertain things.

After all, it wasn't such a very bad shot to land in the vicarage garden, and that practically is what I did. And I don't see that it was my fault that all the vicarage and a lot of friends should be having lunch on the lawn. They were doing that, of course, so as to be on the spot without having to rush out of the house when *Alauda Magna* came home again. Quiet exultation—that was their game. They wanted to gloat over every particular of my ignominious return. You can see that from the way they had arranged the table. I can't help it if Fate decided that my return wasn't to be so ignominious as all that, and swooped me down on the lot of them.

They were having their soup. They had calculated on me for the dessert, I suppose.

To this day I can't understand how it is I didn't kill the vicar. The forward edge of the left wing got him just under the chin and carried him back a dozen yards. He must have had neck vertebrae like steel; and even then I was amazed his head didn't come off. Perhaps he was holding on underneath; but I can't imagine where. If it hadn't been for the fascination of his staring face I think I could have avoided the veranda, but, as it was, that took me by surprise. That was a fair crumple up. The wood must have just rotted away under its green paint; but, anyhow, it and the climbing roses and the shingles above and everything snapped and came down like stage scenery, and I and the engines and the middle part drove clean through the French windows on to the drawing-room floor. It was jolly lucky for me, I think, that the French windows weren't shut. There's no unpleasanter way of getting hurt in the world than flying suddenly through thin window glass; and I think I ought to know. There was a frightful jawbation, but vicar was out of action, that was one good thing. Those deep, sonorous sentences! But perhaps they would have calmed things. . . .

That was the end of *Alauda Magna*, my first airplane. I never even troubled to take her away. I hadn't the heart to. . . .

And then the storm burst.

The idea seems to have been to make mother and me pay for everything that had ever tumbled down or got broken in Mintonchester since the beginning of things. Oh! and for any animal that had ever died a sudden death in the memory of the oldest inhabitant. The tariff ruled high, too. Cows were £25 to 30 and upward; pigs about a pound each, with no reduction for killing a quantity; verandas — verandas were steady at 45 guineas. Dinner services, too, were up, and so were tiling and all branches of the building trade. It seemed to certain persons in Mintonchester, I believe, that an era of unexampled prosperity had dawned upon the place — only limited, in fact, by the solvency of me and mother. The vicar tried the old "sold to defray expenses" racket, but I told him he might sell.

I pleaded defective machinery and the hand of God, did my best to shift the responsibility on to the firm in Blackfriars Road, and, as an additional precaution, filed my petition in bankruptcy. I really hadn't any property in the world, thanks to mother's goodness, except my two motor bicycles, which the brutes took, my photographic darkroom, and a lot of bound books on aeronautics and progress generally. Mother, of course, wasn't responsible. She hadn't lifted a wing.

Well, for all that, disagreeables piled up so heavily on me, what with being shouted after by a ragtag and bobtail of schoolboys and golf caddies and hobbledehoys when I went out of doors, threatened with personal violence by stupid people like old Lupton, who wouldn't understand that a man can't pay what he hasn't got, pestered by the wives of various gentlemen who saw fit to become out-of-works on the strength of alleged injuries, and served with all sorts of silly summonses for all sorts of fancy offences, such as mischievous mischief and manslaughter and willful damage and trespass, that I simply had to go away from Mintonchester to Italy, and leave poor little mother to manage them in her own solid, undemonstrative way. Which she did, I must admit, like a Brick.

They didn't get much out of her, anyhow, but she had to break up our little home at Mintonchester and join me at Arosa, in spite of her dislike of Italian cooking. She found me already a bit of a celebrity because I had made a record, so it seemed, by falling down three separate crevasses on three successive days. But that's another story altogether.

From start to finish I reckon that first airplane cost my mother over £900. If I hadn't put my foot down, and she had stuck to her original intention of paying all the damage, it would have cost her 3,000. . . . But it was worth it. It was worth it. I wish I could live it all over again, and many an old codger like me sits at home now and deplores those happy, vanished, adventurous times, when any lad of spirit was free to fly — and go anywhere — and smash anything — and discuss the question afterwards of just what the damages amounted to and what his legal liability might be.

Terence Hanbury White's first flight was less eventful than that in the preceding story, and so was his ensuing solo. In fact, the peculiar delight of these passages from White's 1936 *England Have My Bones* is that not much "happens" in the way that things do in so many other flying stories—no disasters, near or actual; no violent weather; no failure of vital instruments or equipment. Though not without an overhanging aspect of fear, it is all reasonable and serene. White confronts that fear with the calm and cool sense which we all wish we had mustered in our first flight: "I would recommend a solo flight to all prospective suicides. It tends to make clear the issue of whether one enjoys being alive or not."

T. H. WHITE

Solo

It was a bright moment in a week of suppressed despair. I went to arrange about training at Credon aerodrome, and heard of a newspaper advertisement scheme. They grant you twenty pounds toward your expenses if you are marked by the instructor as the best pupil at his aerodrome.

Then there were many days of trying to find the right coupon to fill in. On Sunday I couldn't wait any more, had sent off the entry forms, rung up the paper for permission to take my lesson before getting their official answer, and fixed a lesson for two-thirty.

There was a little difficulty in swallowing my lunch.

Humphrey took us over in the Lancia. He flew during the war. He said: "Movements almost imperceptible, and you won't feel in contact with your airplane at all during the first hour." It was a piece of luck that he said this, because I had been reading a flying book and had got it into my head that one pushed things about by numbers. What he said put me in exactly the right frame of mind.

I signed a check in the control room with rather a wiggly signature.

When Johnny Burn and the airplane turned up I stopped being nervous.

He said: "I am only allowed to take you up to a thousand and let you fly her straight. I have to report on that."

I was given helmet, goggles; I got in and plugged in and put on the safety belt. We started. The ground went fast, receded. The wind blew down one's back. The horizon was quite hazy to the left; on the right you could just guess it. The trees and fields and the white hangars. Middlehampton like a strawberry rash, reddish and sprawling. Scab? Wen.

Johnny Burn said: "Now I will start the patter. The rudder bars . . . the yawing plane . . . sit sloppily . . . etc. Now I want you to take the rudder. Are you ready? Right. The rudder is yours."

I put the balls of my feet on it like a cat.

"I want you to take a mark and fly straight for it."

I was flying straight. I had the rudder. Humphrey had said I would not feel the controls in contact at all. I just padded with my right foot, a millimeter almost toe-movement or feeling of the muscles, cautiously experimental and delicate. The nose came round an inch. I did the other foot. It worked. I was in contact. The engine was very loud.

Then I flew straight. It was easy, but one felt a little stupefied: Étourdi.

Johnny Burn said: "Good. Now the joystick is more difficult . . . the nose . . . the tail . . . the looping plane. Now I am going to give you the joystick. Are you ready? Right. Now you've got her altogether."

This definitely one wouldn't be able to feel. I touched it with finger and thumb, with a sort of capillary attraction. There was almost a layer of air between my fingers and the stick. We were flying straight. I had got her altogether. We didn't fall out of the air.

I soothed, suggested, fancied, ever so little, over to the left. The wing went down a foot. But it was perfect! We hinted forward, the nose dropped a few inches below the haze. Back, and it came up.

We flew straight.

Johnny Burn said: "Now I shall turn her round. Let me have the controls."

We turned round.

We flew straight.

I looked at the wings now, as well as the nose, and felt them instinctively. It was pie, and very amusing. We were over the aerodrome. Johnny Burn kept on talking.

He said: "If you keep your nose up and lose speed you stall. Let me have the controls. It is like this."

We went up and failed, went soggy. You could feel the air not pressing on the wings.

He said: "Then you stall. Your controls won't work. There is not enough pressure on them."

He waggled everything violently. The ailerons flapped madly, but nothing happened. The nose came down and went toward the earth. It was faintly unpleasant. Then, in the dive, the controls took over again.

Later he switched off the engine and made me take up the right gliding angle. It felt less safe when the engine was off.

He took over.

The S-turns with the engine off felt fragile. The ground was nearer, and there were people on the road, looking up. We swept in over the hedge and the ground was close. We flew along it. You would have hardly noticed when we were on.

I said, getting out: "How many entrants are there for your newspaper scheme?"

He said: "There are ninety. You are number 28, easily the best so far."

This pleased me, because I didn't suspect him then of being a flatterer.

I said: "I could have turned it round myself, I swear, and I want to go up again."

Humphrey wanted to go home. Johnny Burn was booked till six-thirty. I made Ker promise to take me back after tea, by saying I would pay for a lesson for him.

On the way home I sang "*Ai nostri monti*" and the scherzo of Brahms' Quintet of F Minor, neither of them at all in tune. I also bounced on the back seat and talked volubly, until Humphrey looked cross.

We went back after tea.

We went up at once.

Johnny did the patter.

We did banking turns.

It was a different machine, I was in the back seat with the instruments instead of the front one, and the telephone didn't properly work. I missed most of what he said. But it didn't matter a scrap. Flying is the easiest and the most amusing thing in the world. You don't fly by rules or numbers, but by instinct and by feeling the pressure of the wind on your wings. I didn't have to think about anything except my rudder bar, which is the only slightly unnatural movement. I was allowed to play with the engine, climb, glide, dive, and bank. Also with the tail trimming gear. I have had fifty minutes in all. Aviators live by hours, not by days. . . .

JUNE 15, 1934.

Well, well, well, well, well. Yesterday I had to do some work, so it was impossible. Today I got over before ten-thirty, only to find that all the machines were out of action. AV's petrol tank was leaking, ET was not insured, and JT had gone off to Carlands for a top overhaul. It was misty, and Richmond had evidently decided that he would lose his way if he came back. So we hung about the bar, and played with the gambling machines, and generally groused till luncheon.

I am beginning this at the wrong end. On the drive over I tried to reckon out my attitude toward solos. Subconsciously last time I must have been averse from doing it, and I was certainly terrified. If I had been able to go off in AV I think I should have done it, solely because finding myself alone in the air would have frightened me into my senses. But driving over today I was not scared. There was the same intellectual desire to do it, and not the emotional negative. Going solo is rather like committing suicide. One's brain, which must always advocate the suicide's grave as the only reasonable solution, finds itself in conflict with one's vitality. I would recommend a solo flight to all prospective suicides. It tends to make clear the issue of whether one enjoys being alive or not.

Driving over, I knew that I wanted to do it, basically liked being alive; and this time was going to be less scared. I had thought at the end of last time that I should be absolutely unscared. No. I was nervous when I actually thought about it from close to.

I was flagged by a tramp of about sixty-five and stopped for him. He wanted to go to Middlehampton: said he was glad I had stopped, because he would never have got there: was "beginning to feel *languid*." He had walked all night from Warden. It was a warm night, he said, and, proudly, he had not needed to put on his overcoat. He had been flagging people since 8 A.M., but none of them had stopped. I explained apologetically about motor bandits, though anything less like a bandit than this frail, rosy-cheeked old gentleman with his careful overcoat it is difficult to conceive. When I put him down I apologized for lack of money. Oddly enough, it was true that I had not even a penny, as I was in my best suit and had forgotten to change over the cash. He had not asked, and with an unoffended dignity denied a desire for any. He thanked me for what I had done.

Then no machines till luncheon. Then a very good lunch indeed of cold meat, salad, fruit and cream, gorgonzola.

Then Johnny said: "Off we go."

It was JT (no slots) and little wind. What there was came from the east, necessitating a long hold off. We did thirty minutes.

Johnny was in a better temper and I had told him at lunch that he ought to be more patient with me. The amazing result of this was that he was. I was landing badly — bouncing all over the place, swinging my tail, and not getting to exactly the right mark for my glide in. Johnny kept having to slip off height for me. Twice, I think, I had to throttle on and fly round again. But I had mastered JT's cheese-cutter for the glide. I did one three-pointer.

Johnny said: "Two more, and you go."

The next circuit was passable: a wheel landing, but effective for its purpose.

The next circuit was a simply terrible bounce.

Johnny said nothing.

I watched the back of his head as we taxied back to the takeoff. It looked undecided.

He put her in position and climbed out.

He said: "Righto. Don't forget she will climb quicker without me, and glide flatter. For God's sake throttle on and go round again if you are going to under or overshoot. You don't need to come down the first time. There's petrol for half an hour."

I was not so frightened as before, but still monosyllabic. I said: "Yes."

Johnny said: "I shall just wait about. Good-bye."

I said: "Good-bye."

Well, I thought, one can begin, at any rate. Life had become a thing which only existed in fractions of seconds ahead.

I took off, beautifully it seemed. When I was two feet up I thought: Now it's too late to land.

She climbed buoyantly, feeling much more carefree and speedy, rising like a balloon. Three hundred feet. I turned at the road. I thought: "Am I frightened?"

Yes.

I thought it would be well to become automatic again, as I had been for the last thirty minutes, instead of going in for feelings.

At the house I turned again, but we were over 400.

I wanted to reach a ride in the northerly wood at 400, bank in along it, switch off diagonally to the field, and come in along the other diagonal, over a red barn.

I was too high, from being alone.

I hurried downstairs: not by switching off and gliding, but by putting the nose down with engine on. This made us go very much faster, which I didn't like. Nor, he said afterwards, did Johnny, a lonely figure in the middle of the aerodrome.

I got the revs exactly where I wanted, but found I was frightened of flying fast.

Then there was a big bump. Do bumps bump more when one is alone? I didn't like it. But one was too busy to feel much about anything.

There was the ride. We spun over sideways, swept along it,* and switched off.

Back came the cheese cutter. It was a lovely, gentle glide.

I did my gliding turn, beginning to take it off long before it was over. Fine.

To my mild surprise I was pointing exactly for the barn and along my diagonal.

Then I saw that I was going to undershoot. I was going to hit the barn itself with my undercarriage. I kept my hand on the throttle, thinking: "Give her time, it is not sure, we may just scrape in." Anyway, it was not much worse hitting a barn than the earth. I felt I could judge it to a foot, and had time to switch on at a tenth of a second. We were going to make it.

We did. The tail was clear by several feet.

I began to flatten out, to flatten. She was not swinging or yawing and neither of her wings were down. My feet on the rudder bars were perfect.

I was going to come down alive.

We flattened, flattened, straight, perfect, touching, touching, tail, rumble, still straight, even on the ground, and not a bounce. Not a bounce.

He had done his solo.

He shook hands with himself.

He tried to be monosyllabic with Johnny, but was talking fast, because he was still living very fast. Johnny got on the wing and the step. He taxied him back.

Johnny said: "I thought you were going to switch on and go round again."

There were very few people in the club. They came out like disturbed ants. A girl friend, Ian, Richmond.

They shook hands with Tim — a solemn moment. Then they drank some lager.

In the logbook "Solo" is written in red ink.

*In practically a racing turn, because of the extra speed I had gathered from my descent with engine on. It must have looked very funny from the ground.

Following upon three years of increasingly successful glider flights, the Wrights left their native Dayton, Ohio, in the fall of 1903, as they had done before, for the windy beaches of Kitty Hawk, North Carolina. This time, however, they brought along a gasoline engine of their own design and construction; they would attempt manned, powered flight. Their initial venture was to have come on December 14, only six days after Samuel Pierpont Langley's *Aerodrome* had plunked into the Potomac for the second time that year. The Wrights' December 14 attempt came to naught, as "pilot error" by Orville aborted the flight almost as it began, in the process breaking the rudder. The rudder repaired, three days later came the culmination of the Wrights' dream, and man's. Here is the report of that momentous event, from the diary of Orville Wright.

ORVILLE WRIGHT

December 17, 1903

When we got up a wind of between twenty and twenty-five miles was blowing from the north. We got the machine out early and put out the signal for the men at the [life saving] station. . . . After running the engine and propellers a few minutes to get them in working order, I got on the machine at 10:35 for the first trial. The wind at this time was blowing a little over twenty-seven miles. . . . On slipping the rope the machine started off increasing in speed to probably seven or eight miles. The machine lifted from the truck just as it was entering on the fourth rail. Mr. Daniels took a picture just as it left the tracks. I found the control of the front rudder quite difficult on account of its being balanced too near the center and thus had a tendency to turn itself when started so that the rudder was turned too far on one side and then too far on the other. As a result the machine would rise suddenly to about 10 feet and then as suddenly, on turning the rudder, dart for the ground. A sudden dart when out about 100 feet from the end of the tracks ended the flight. Time about 12 seconds. . . . At just twelve o'clock Will started on the fourth and last trip. The machine started off with its ups and downs as it had before, but by the time he had gone over 300 or 400 feet he had it under much better control, and was traveling on a fairly even course. It proceeded in this manner till it reached a small hummock out about 800 feet from the starting ways, when it began its pitching again and suddenly darted into the ground. . . . The distance over the ground was 852 feet in 59 seconds.

Wilbur & Orville WRiGHT

On December 17, 1903, at Kitty Hawk, N.C. . . .

MAN FLEW IN A HEAVIER-THAN-AIR MACHINE FOR THE FIRST TIME.

ORVILLE, ON A COIN TOSS, SERVED AS THE PIONEER PILOT.

HIS HISTORIC FLIGHT LASTED **12** SECONDS AND COVERED **120** FEET.

On September 18, 1901, Wilbur Wright, whose glider experiments with his brother Orville had aroused much interest, came to Chicago to address the Western Society of Engineers. These words, taken from that address titled "Some Aeronautical Experiments," form an appropriate envoi for armchair aviators.

WILBUR WRIGHT

A Fractious Horse

Now, there are two ways of learning to ride a fractious horse: one is to get on him and learn by actual practice how each motion and trick may be best met; the other is to sit on a fence and watch the beast a while and then retire to the house and at leisure figure out the best way of overcoming his jumps and kicks. The latter system is the safer, but the former, on the whole, turns out the larger proportion of good riders. It is very much the same in learning to ride a flying machine; if you are looking for perfect safety, you will do well to sit on a fence and watch the birds, but if you really wish to learn, you must mount a machine and become acquainted with its tricks by actual trial.

A poem of World War I, by a titan.

WILLIAM BUTLER YEATS

An Irish Airman Foresees His Death

I know that I shall meet my fate
Somewhere among the clouds above;
Those that I fight I do not hate,
Those that I guard I do not love;
My country is Kiltartan Cross,
My countrymen Kiltartan's poor,
No likely end could bring them loss
Or leave them happier than before.
Nor law, nor duty bade me fight,
Nor public men, nor cheering crowds,
A lonely impulse of delight
Drove to this tumult in the clouds;
I balanced all, brought all to mind,
The years to come seem waste of breath,
A waste of breath the years behind
In balance with this life, this death.

In an odd coincidence, the last piece in this book alphabetically is also the last chronologically. On April 12, 1981, space shuttle commander John Young and pilot Robert Crippen embarked upon an adventure which opened a new era of space travel: the *Columbia* orbiter and its twin rocket boosters, enormously expensive though they were, would return to earth more or less intact and be reusable, unlike all the spacecraft of the previous twenty years. Surely this would be the road to future industrialization—and, alas, militarization—of space. Here is the story of the shuttle's maiden voyage, in the words of the astronauts.

JOHN YOUNG and ROBERT CRIPPEN

Our Phenomenal First Flight

Young: We had a complex flight plan, detailing what we were to do almost minute by minute. After we finished the first day's chores, Crip fixed us dinner. Mission Control told us it was bedtime and signed off for the next eight hours. Neither one of us slept well that first night. For one thing it was light out much of the time and far too beautiful looking down at earth.

I had taken along a 70-mm Hasselblad camera. Before launching, some geologists, oceanographers, and meteorologists I know had told me things they would like to see from space. Things like dune patterns, evidence of internal waves moving below the ocean surface, sediments off mouths of rivers, and clouds forming near coastlines and islands. Most of the time we were too busy to take pictures on this mission, but it was so easy to do after working hours.

Sixteen years ago on *Gemini 3* we didn't have any windows to speak of. There was one porthole in front of me and one in front of Gus Grissom. The only way we could take pictures was to point a camera straight at something or open the hatch. (We didn't do that much.) On *Apollo* we were on our way to the moon. We didn't have much chance to look back and take pictures. We were moving too fast anyway.

Crippen: The shuttle has those wraparound windows up front. But the best views are from the flight deck windows, looking out through the payload bay when you are flying upside down with the doors open, which we were doing most of the time. You see the whole earth going by beneath you.

I remember one time glancing out and there were the Himalayas, rugged, snow covered, and stark. They are usually obscured by clouds, but this day was clear and the atmosphere so thin around them that we could see incredible detail and vivid color contrast. The human eye gives you a 3-D effect no camera can. Sights like the Himalayas and thunderstorms, which we later saw billowing high above the Amazon, are especially dramatic.

John and I have spent a lot of time with the navy in the Mediterranean. It often has a haze over it. When you are flying over the Med at night and you see ships out on the sea below and stars up above, you can lose all sense of what is up and what is down. Luckily the Med was fairly clear for us now. We had gorgeous views of Gibraltar, the Sahara, and the Bay of Naples. We could see Mount Etna smoking. Whenever the sun was setting, the sun glint on the water let us see ships on the surface hundreds of miles away. Perhaps the most stunning sight, however, was Dasht-e Kavir, a salt desert in Iran. It looked more like Jupiter with the great swirls of reds and browns and whites, the brilliant residues of generation after generation of evaporated salt lakes.

Young: I really liked the Bahamas. They glowed like emeralds. Unfortunately the pictures could not capture their shimmering beauty. The human eye is so much better for seeing colors and contrasts. The human being, with the detail he can pick out, will prove to be very useful in space.

I would like to get together with our photographic guys and touch up the colors in our pictures. That would probably not be cricket, but it is disappointing to look at Eleuthera and not see that emerald glowing.

I wasn't ready to go to bed that first night at quitting time even though we had been up for eighteen hours. I slept only three or four hours. Crip did a bit better. When we did turn in, we just fastened our lap belts and folded our arms. We could have gone down to the middeck and just floated around, but I like some support. Any way you do it, sleeping in zero G is delightful. It is like being on a water bed in three dimensions.

Crippen: We were busy most of our second day, April 13, doing burns with the reaction control jets, going into different attitudes and performing maneuvers. We needed to understand how well the computer autopilot can control the vehicle. Could we make fine maneuvers? Houston wanted to see how well the crew could coordinate with the ground in positioning the orbiter.

Young: I just kept feeling better and better about that vehicle. After we launched and got it into orbit, I had said to myself, "Well, that went pretty good." Then the vehicle worked so well the first day I had said, "We'd better take it back before it breaks." The second day it worked even better and so I thought, "Man, this thing is really good. We'd better stay up here some more to get more data." But Mission Control made us come back the next day.

Crippen: We both slept soundly that second night. I was really sawing the z's when an alarm started going off in my ears. I didn't know where I was, who I was, or what I was doing for the longest time. I could hear John saying, "Crip, what's that?" It was a minor problem, fortunately. A heater control in one of our auxiliary power units quit working. We just switched on an alternate heater and went back to sleep.

It was about 2:30 A.M. Houston time when flight control greeted us with a bugle call and some rousing music. John fixed breakfast that morning, although usually I took care of the chow. Then we checked out the flight control system one last time and stowed everything away for reentry. We strapped on biomedical sensors to keep the doctors happy, and got back into our pressure suits. We programmed the computers for reentry and closed the payload bay doors.

The first step toward getting home was to deorbit. We had tested all our engines and were very confident they were working. We were really looking forward to flying reentry. Bringing a winged vehicle down from almost twenty-five times the speed of sound would be a thrill for any pilot.

We were orbiting tail first and upside down. We fired the OMS engines enough to feel a nice little push that slowed us down by a little less than 300 feet per second. That is not dramatic, but it did change our orbit back to an ellipse whose low point would be close to the surface of the earth.

When we finished the OMS burn, John pitched the vehicle over so it was in the 40-degree nose-up angle that would let our insulated underbelly meet the reentry heat of the atmosphere.

Young: We hit the atmosphere at the equivalent of about Mach 24.5 after passing Guam. About the same time we lost radio contact with Houston. There were no tracking stations in that part of the Pacific. Also, the heat of reentry would block radio communications for the next sixteen minutes.

Just before losing contact, we noticed a slight crackling on the radio. Then, out of the sides of our eyes, we saw little blips of orange. We knew we had met the atmosphere. Those blips were the reaction control jets firing. In space we never noticed those rear jets because there were no molecules to reflect their light forward. Those blips told us that *Columbia* was coming through air — and hence plenty of molecules to reflect the thrusters' fire.

That air was also creating friction and heating *Columbia*'s exterior. About five minutes after we lost contact with Houston, at the beginning of reentry heating, when we were still flying at Mach 24.5, we noticed the reddish pink glow. Bob and I put our visors down. That sealed our pressure suits so that they would automatically inflate if somehow reentry heating burned through the cabin and let the air out. Other than the pink glow, however, we had no sense of going through a hot phase.

Crippen: *Columbia* was flying smoother than any airliner. Not a ripple!

As we approached the coast of northern California, we were doing Mach 7, and I could pick out Monterey Bay. We were about to enter the most uncertain part of our flight. Up to this point, *Columbia*'s course was controlled largely by firings of its reaction control thrusters. But as the atmosphere grew denser, the thrusters became less effective. *Columbia*'s aerodynamic controls, such as its elevons and rudder, began to take over.

We had more and more air building up on the vehicle, and we were going far faster than a winged vehicle had ever flown. Moreover, the thrusters were still firing. It was an approach with a lot of unknowns. Wind tunnels just cannot test such complex aerodynamics well. That was the main reason John took control of the flight from the automated system at a little under Mach 5. We had been doing rolls, using them a little like a skier uses turns to slow and control descent down a mountain. The flight plan called for John to fly the last two rolls manually. He would fly them more smoothly than the automatic system, helping to avoid excessive sideslipping and ensuring that we would not lose control as we came down the middle of our approach corridor.

Young: It turned out to be totally unnecessary for me to manually fly those last two roll reversals. *Columbia* had been flying like a champ. It has all those sensors: platforms for attitude control, gyroscopes, and accelerometers. Its computers take all the data, assimilate it instantly, and use it to fire thrusters, drive elevons, or do anything needed to fly the vehicle. They are much faster at this than any man. The orbiter is a joy to fly. It does what you tell it to, even in very unstable regions. All I had to do was say, "I want to roll right," or "Put my nose here," and it did it. The vehicle went where I wanted it, and it stayed there until I moved the control stick to put it somewhere else.

Crippen: Flying down the San Joaquin Valley exhilarated me. What a way to come to California! Visibility was perfect. Given some airspeed and altitude information, we could have landed visually.

John did his last roll reversal at Mach 2.6. The thrusters had stopped firing by then, and we shifted into an all-aerodynamic mode. We found out later that we had made a double sonic boom as we slowed below the speed of sound. We made a gliding circle over our landing site, Runway 23 on Rogers Dry Lake at Edwards Air Force Base.

On final approach I was reading out the airspeeds to John so he wouldn't have to scan the instruments as closely. *Columbia* almost floated in. John only had to make minor adjustments in pitch. We were targeted to touch down at 185 knots, and the very moment I called out 185, I felt us touch down. I have never been in any flying vehicle that landed more smoothly. If you can imagine the smoothest landing you've ever had in an airliner, ours was at least that good. John really greased it in.

"Welcome home, *Columbia*," said Houston. "Beautiful, Beautiful."